W9-BWS-674

Introduction to

Geographic Information Systems
and *Remote Sensing*

Concepts

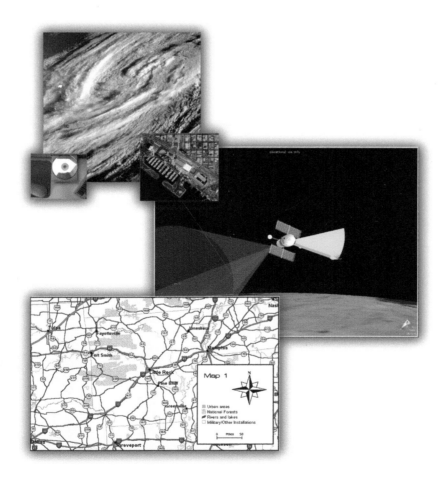

Statement of Copyright

Introduction to GIS & Remote Sensing Concepts

Table of Contents - Student Manual

Unit 1: Fundamentals of Geospatial Technology

An overview into the four main areas of Geospatial Technology. Unit Features a brief narrative of each technology and an exercise applying the concepts. The final lesson is an exercise combining all areas.

Unit 2: Introduction to Project Management

An introduction into project management methods and a project management framework – the STARS Project Management Model. Each lesson features detailed narratives of each phase, an exercise applying the concepts, and a lesson review to reinforce key concepts.

Unit 3: Introduction to GIS Concepts

An introduction into the major concepts of mapping and Geographic Information Systems including histories, coordinate systems, projections, and key analysis concepts. Each lesson features detailed narratives of each phase, an exercise applying the concepts, and a lesson review to reinforce key concepts.

Table of Contents

Preface

This introductory course launches you into the exciting world of Geographic Information Systems and Remote Sensing. While learning about the basics from the evolution of maps and projections to learning about the modern uses of a GPS, GIS, and Remote Sensing, you will complete many "hands-on" activities such as creating your own maps using computers, rulers, and tape measures. You will also utilize an actual program that NASA uses to simulate satellite movements. The specific areas of focus for this course will be an Introduction to GIS and Remote Sensing, an Introduction to the Project Management Model, an Introduction to GIS concepts, an Introduction to Remote Sensing Concepts, and the Satellite Tool Kit. Digital Quest's goal is to combine our experience in geospatial technology and unique knowledge of GIS/RS education with industry professionals and institutions to bring this technology to future geospatial technicians and students. This book is intended for those who want to gain basic geospatial skills and learn the concepts driving these skills as well as the skills of tomorrow. We created a manual that will not only introduce the "buttonology" of how to use geospatial software but also to provide an understanding of those processes that can be used in the professional or educational setting. This curriculum utilizes real-world data to simulate geospatial scenarios and applications. Geospatial technology is enhancing all manners of life and industry, and we want you to have a working knowledge of these tools.

This book is designed for secondary, post secondary, and professionals. There are neither prerequisites nor is previous GIS experience necessary. We provide all directions and data needed to complete the directed tasks.

How will I use this book?
This book is designed to be completed sequentially with lessons that provide an overview of all geospatial technologies with subsequent lessons examining GPS, Remote Sensing, GPS, and Aerospace Technology and its importance to Geospatial Technologies. Each lesson will focus on a necessary concept that is fundamental to a users growth as geospatial technician.

Features in this book:

- **Step-by-Step Instructions** –an easy to follow format relevant for novice to experienced ArcGIS users
- **Lesson Content** – provides a narrative overview of an important topic relevant to the fundamental concepts of geospatial technology
- **Lesson Exercise** – applies lesson content to an exercise that features "real world" tasks at work
- **Lesson Review** – reinforces knowledge gained with exercises to identify key terms and concepts
- **"Knowledge Knugget" boxes** – boxes found in the margin that features tools, tips and tricks that may enhance your experience with geospatial technologies
- **Finished Layouts** – to provide a self assessment tool to ensure successful completion of each lesson

About ESRI's ArcGIS

ArcGIS is a software suite developed by Environmental Services Research Institute, Inc. (ESRI) designed to analyze and model geospatial data. Of the software suite in ArcGIS, this book will use three major components; ArcMap, ArcToolbox and ArcCatalog. ArcMap is the primary part of the suite that will be used throughout the book to display, create, and analyze different types of geospatial data. ArcToolbox contains various geoprocessing tools used throughout the ArcGIS suite to complete various tasks such as creating buffers, merging shapefiles, and address locators. ArcCatalog is the "virtual filing cabinet" where users create, manipulate, or preview data and metadata. ArcGIS is widely accepted and used among today's GIS professionals and students. Using this software will make for a smooth transition for the student to take their GIS skills from the classroom to the workplace or other academic pursuits.

Teacher Materials

The STARS Introduction to Geographic Information Systems and Remote Sensing Concepts is designed to be used in a variety of learning environments. For classroom environments, a teacher's edition is available with the following enhancements:

Overviews – Each lesson in the teacher's edition comes with a lesson overview page for instructors with boxes in the margins that provide quick reference to lesson goals. The "What Will You Teach" section provides the instructor with a bulleted list that includes a list of goals for the lesson GIS skills learned. The "How Will You Teach It" section provides the instructor with the procedures to introduce the topic and have the students complete the lesson. The initial introduction provides a unique perspective on the topic that is covered in the lesson that will enhance teacher's abilities to facilitate learning in a classroom environment. The overview also describes in detail the skills taught in the lesson as well as additional information that may be necessary to complete the lesson. Lesson related links for further study are also supplied with each overview page.

PowerPoint Presentation Notes – Each student lesson comes with a PowerPoint presentation that provides an overview of the lesson including concepts, the skills they will cover, and the study area involved. The teacher's manual is supplemented with detailed descriptions and commentary for each slide allowing a diverse range of instructors to lead classroom lecture. For every presentation, the student manuals will have notes sheets, complete with thumbnail pictures of the slides from the presentation, with lines for notes beside them.

Assessments – Lessons will conclude with a full page color layout of a successfully completed exercise. If questions are presented within a lesson, the teacher's manual includes answers to those questions.

About the Authors

Eddie Hanebuth is founder and president of Digital Quest, a Mississippi-based development and training-oriented company that produces GIS instructional material for educational institutions. He chairs the U.S. Department of Labor's National Standard Geospatial Apprenticeship Program and the SkillsUSA Geospatial Competition Committee, and runs the SPACESTARS teacher-training laboratory in the Center of Geospatial Excellence, NASA's John C. Stennis Space Center.

Liz Rotzler has six years experience in geospatial technology education. After teaching GIS in the classroom using the STARS curriculum and certification, she has spent the past three years working in development of GIS/RS curriculum. She has co-authored and edited the first book in Digital Quest's aGIS series, Introduction to Geospatial Technologies as well as worked in the popular Digital Quest SPACESTARS series.

Austin Smith has been part of the Digital Quest team for four years. He is currently the Vice President of Development and Support and also serves as the chair of the S.T.A.R.S. Geospatial Certification Committee. He has experience in information technology development, implementation, and training in a variety of public and private organizations. At Digital Quest he has co-authored or edited over 30 titles. With Digital Quest's STARS series, he serves in authoring, planning, and final editing.

Fundamentals of Geospatial Technology

Learning Topics

Explore the Geospatial Technologies:
o GIS
o GPS
o RS
o Surveying

Discover why Geospatial Technology is important

Introduction to Geographic Information Systems and Remote Sensing *Concepts*

Unit One

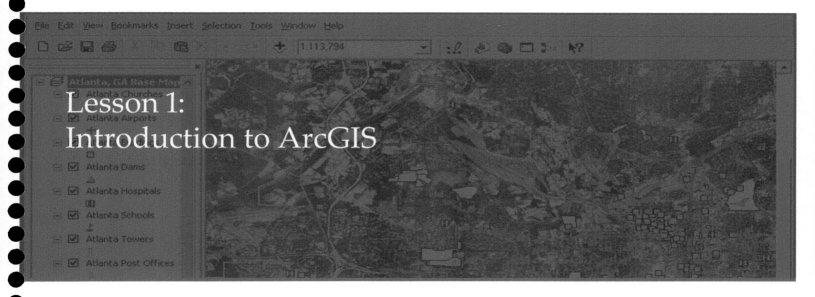

Lesson 1:
Introduction to ArcGIS

When visiting a place that you have never before visited, you have several options; use a street map, use a computer with an internet program, or use a navigational device – whether it be free standing, built into your automobile, or even your cell phone. Each method has its pros and cons, but in the end, you want a reliable source that will get you to your destination.

Atlases, road maps and street maps have been a reliable means for finding locations for years. They come complete with an index and a scale to provide you with an idea of how far away a destination is from a starting point. You find out where something is, try to find the best route, and then calculate how long it takes to travel there. With newer technology, finding out where another place is and how long it will take to get there, requires much less effort. With a few clicks of your mouse you can find the answer to any traveling question. What type of driving force is behind this type of technology? Geospatial technology. If you have used a device such as an online mapping program or a navigational device (even those in cell phones), the technology has already touched your life.

Although the world has been operating for many years without it, geospatial technology is making our lives easier by allowing us to not only get information quickly but also allows us to "see" the answers to so many questions that we have. This lesson will introduce you to the most widely used geospatial software, ArcGIS.

Introduction to ArcMap

1. ***Start*** ArcMap by ***double-clicking*** the 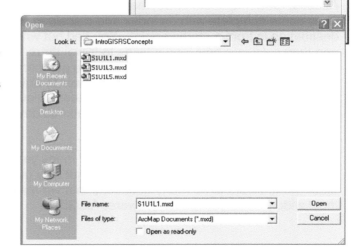 shortcut on your computer desktop (or by ***clicking*** the Windows Start button **start** , ***pointing*** to **All Programs** ▷ ArcGIS ▶ and ***selecting*** ArcMap .)

2. When the ArcMap window appears, make sure that the radio button next to the (Start using ArcMap with) ⊙ An existing map: option is selected.

    At the bottom of the ArcMap startup dialog box, make sure that Browse for maps... is highlighted.

3. ***Click*** OK .

4. In the **Open** dialog box, ***navigate*** to the **C:\STARS\IntroGISRSConcepts** folder.

5. ***Select*** the **S1U1L1.mxd** file.

6. ***Click*** Open .

A **map document** will open displaying major cities, roads, rivers, state boundaries, and counties layers for the 48 contiguous United States. This is a file that has been specifically created for this lesson. Throughout this lesson, you will be introduced to different ways of navigating through the data supplied. This will be the first map that you will save to your student folder.

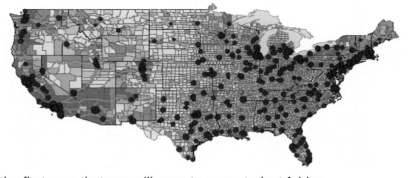

---

7.   To save this ArcMap map document in your student folder, **select Save As…** from the **File** menu.

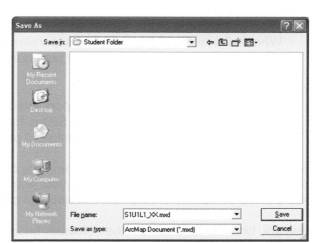

8.   **Navigate** to a location specified for you to save your work on your computer hard drive. **Name** the new file **S1U1L1_XX** (where **XX** is your initials) and **click** Save .

Why is geospatial technology useful?
Using geospatial technology can help you answer geographic questions like...
- Where is Houston?
- What is the name of the largest county in the contiguous United States in area?
- What is the name of the river that separates Minnesota from Canada?
- Which states use water for either a boundary or a portion of their boundary?

Or answer demographic questions like...
- How many people live in the most populated county in the United States?
- What city in the contiguous United States has the highest population?
- Which state(s) have the fewest number of people per square mile?

Geospatial technology is not limited to demographic studies and includes many disciplines from biology to political science. As you work your way through this manual, you will discover many uses and will be able to create future projects to cater to your needs.

Once you become familiar with the software, finding answers to questions such as those posted above will not only be easier, but you will also discover, in most cases, there is usually more than one way to find the answer you are looking for.

Answering the question, "Where in the United States is...?" using ArcGIS
One great benefit of using software like ArcMap is that you are able to "see" the answer to just about any question.

- Where is Houston?

Most people know that Houston is a rather large metropolitan area in Texas. But where in Texas is it...accuracy is essential for proper analysis? Is it in the center? Near the coast? In North Texas? In South Texas? Let's use ArcMap to find the answer.

1.   In the **ArcMap** window, ***right click*** on the ☑ Cities layer.

2.   From the menu that appears, ***select*** 🔳 Open Attribute Table   . A table will appear. The fourth column will contain the cities listed in the table. These are also the cities that are shown on the map.

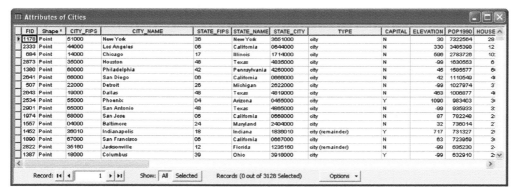

Instead of scrolling through the list of over 3000 city names, you will alphabetize them to make it easier to find Houston in the list.

3.   *Right click* on the field heading labeled **CITY_NAME**.

4.   *Select* ≜ Sort Ascending   from the context menu.

5.   *Using* the **scroll bar** on the right side of the table window, *maneuver* down the list until you find Houston.

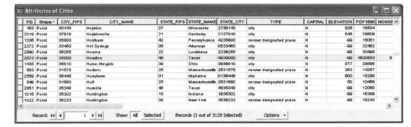

6.   *Click* on the **gray** box ☐ to the left of Houston to select this row.

7.   *Minimize* ◻ the **Attributes of Cities** table so that you will be able to see the map display. Houston will be *highlighted* in the map display.

8.   *Right click* on ☑ Cities and *select*

        Selection   ▶   🔲 Zoom To Selected Features   .

      Your map will now zoom in to the Houston area.

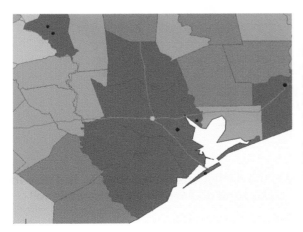

Houston is located near the coast of Texas. Can you think of other questions that could be explored in this area? (i.e. What is the population of Houston and the surrounding areas? Which two interstates cross in Houston? How far is Houston from the coast? All of these questions and more can be answered from this map and its data layers.)

9.   *Single click* the **Go back to Previous Extent** ⬅ button.

10.   *Single click* the **Clear Selected Features** 🔲 button on the Tools toolbar to *deselect* Houston. This will clear the selected entry in both the table and on the map.

Answering the question, "What is the largest area...?" using ArcGIS

- What is the name of the largest county in the contiguous United States in area?

Think you know the answer to this question? Let's use ArcMap to find out if you are correct.

11.   **Right click** on the ☑ Counties layer and **select** ▦ Open Attribute Table .

In order to find the name of this county, you will have to sort the data in the table by area. The easiest way to sort them is in Descending order so that the county with the largest area will appear at the top of the list.

12.   **Right click** on └─ AREA ─┘ and **select** ☰ Sort Descending .

13.   **Single click** on the **Gray box** on the left of the table to **select** the first entry.

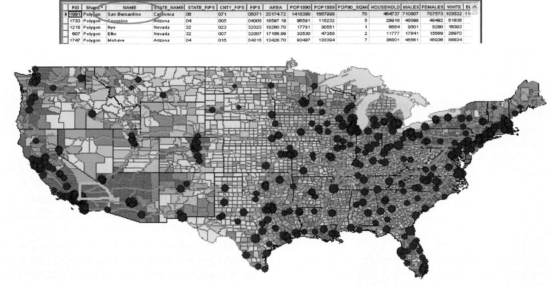

You have now located the largest county in the contiguous United States. Is it also the county with the highest population? (Hint: Just one additional sort in this table will answer that question for you.)

14.   **Single click** the **Clear Selected Features** ⊠ button on the Tools toolbar to **deselect** the selected county.

How do I Find and Identify Specific Features Using ArcGIS?

- What is the name of the river that separates Minnesota from Canada?

Two other selection tools are the Find Tool and the Identify Tool. Both are similar in what they do, but each has a unique feature that makes it more appealing for certain situations. To understand their unique qualities, let's use them to find the answer to this question.

To determine this river, you first need to determine which state is Minnesota.

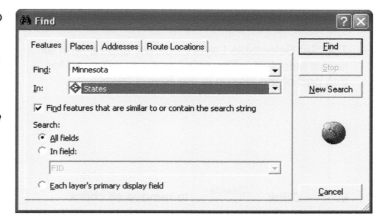

1. *Single click* the **Find** tool 🔍 . The **Find** tool dialog box will appear.

2. In the **Find** tool dialog box, *type* **Minnesota** in the **Find** box.

3. *Click* on the **down arrow** ▼ in the **In** box and *select* **States**.

4. *Click* [**Find**] to start the process. The box will expand, showing the result at the bottom.

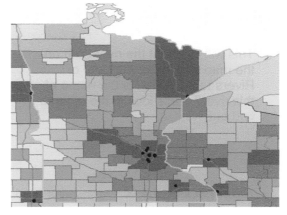

5. *Right click* on the entry and *choose* **Select**. The state of Minnesota will be selected on the map.

6. Use the **Zoom** tool 🔍 to zoom in to this state.

7. Now that you are zoomed in, *single click* the **Clear Selected Features** ⬚ button to clear Minnesota.

You will now use the Identify tool to identify the river at the northern border of this state.

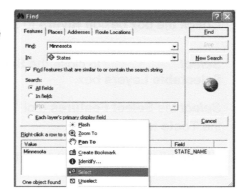

Identifying a Map Item Using the Select Features Tool

Before using the Identify tool, it is best to set the rivers layer so that its features can be selected interactively with the Select Features tool.

8. From the **Main Menu**, *click* Selection, then **Set Selectable Layers**.

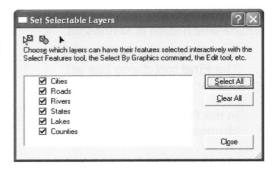

The Set Selectable Layers box will appear with all layers selected.   The only layer you will need for this part of the lesson is the Rivers layer.

9. ***Click*** Clear All to remove all check marks.

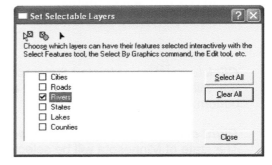

10. ***Check*** the box beside **Rivers** to select it.

11. ***Select*** Close to close the **Set Selectable Layers** box.

12. ***Select*** the **Select Features** tool ⟋ from the **Tools** toolbar.

13. ***Single click*** on the river to select it.

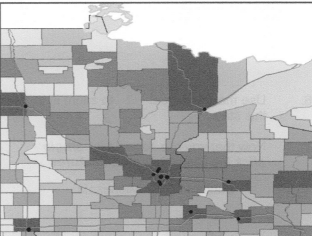

This will highlight the river both on the map and in the Attribute of Rivers table. What is the name of this river? Let's look in the table to find out.

14. ***Right click*** on the ☑ Rivers layer and ***select* Open Attribute Table**.

15. At the bottom of this box, **click** 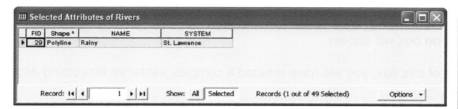 Selected to show the one record that is selected in this table. The name of the River and its System will be shown.

> Do you have to click Selected to see the answer? No. It can be a time saver to avoid having to scroll through the table to find the highlighted item.

Using Select by Location to Answer a Question in ArcMap

- Which states have water for either a boundary or a portion of their boundary?

The borders of many states, counties, countries, and other political features are delineated using physical boundaries like rivers or waterways. To see this on your ArcMap document only takes a few clicks.

1. **Close** the **Selected Attributes of Rivers** table by **clicking** on the ⊠.

2. **Click** the **Clear Selected Features** button ⊠ on the **Tools** toolbar.

3. **Single click** the **Full Extent** ● button.

4. **Turn off** the **Counties**, **Roads** and **Cities** layers by unchecking them on the left side of the ArcMap window (also known as the Table of Contents).

   Your table of contents and map window should look like this:

How many states use rivers or waterbodies as part of its boundary? Although an entire side or border of a state may not be a river segment, there are numerous instances where a small part of the state is bordered by water. Exactly how many can be determined by using Select by Location.

5.   From the **Main menu**, *click* **Selection** and then 🖳 Select By Location...   . The **Select by Location** box will appear.

Upon completion of this box, you will have created a complex sentence instructing ArcMap what to look for.

6.   In the first box, *verify* that **select features from** is selected.

7.   *Check* the **States** layer in the second box.

8.   *Change* the third box to **share a line segment with**.

9.   *Select* Rivers as the feature in the fourth box.

10.   *Leave* the buffer box open.
The sentence should read:
          I want to select features from the following layer, States that share a line segment with the features of this layer, Rivers.

11.   *Click* ⸻ OK ⸻ to perform this query.

All of the states with water as a part of their border or borders will appear highlighted in the map. Instead of counting them, use the attribute table to find your answer.

12.   **Right click** on the

☑ States layer in the table of contents and *select* **Open Attribute Table**.

          The answer to the question is 38 states!

Why are there 49 states listed? The District of Columbia is listed as its own state.

13.   Verify the answer by looking at the bottom of the Attributes of States table.

          Records (38 out of 49 Selected)

14.   *Click* the **Clear Selected Features** button 🔲 on the **Tools** toolbar.

Using Select by Attributes to Query ArcMap …
Up to this point, you have used ArcMap to answer geographic questions. ArcMap is also useful for answering demographic questions.

- How many people live in the most populated county in the United States?

Over 9,000,000 people reside in this popular place. Let's use another search method to determine where this county is.

1. ***Turn on*** the **Counties** layer in the Table of Contents by ***checking*** it. The counties will redraw on the map.

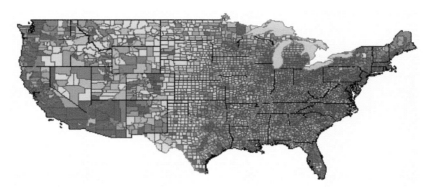

2. From the **Main Menu**, *select* **Selection** and 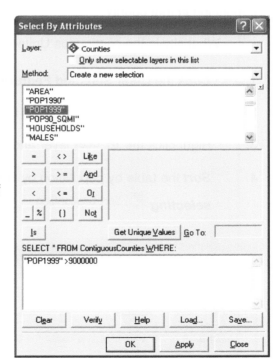 Select By Attributes… . The **Select By Attributes** box will appear.

3. ***Change*** the **Layer** to ◈ Counties by clicking on the down arrow ▼ to the right of the box and ***selecting*** **Counties** from the list provided.

4. ***Double click*** on "**POP1999**" to make it appear in the bottom window.

5. ***Single click*** on the > to make it appear next to POP1999 in the bottom window.

6. ***Type* 9000000** (without commas) to the right of the > sign.

    The Select By Attributes box should look like this:

7. ***Click*** ___OK___ to run the query.

The most populated county will appear highlighted on the map and will also be highlighted in the Counties Attribute table.

8.   ***Right click*** on ☑ Counties in the Table of Contents and ***select* Open Attribute Table** from the context menu.

9.   At the bottom of the **Attributes of Counties** box, ***click*** Selected to show the one record that is selected.

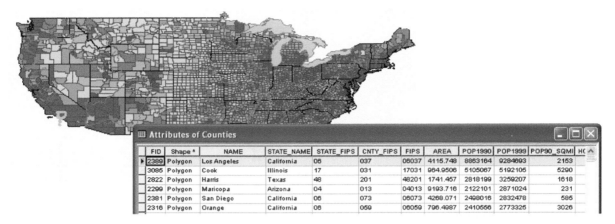

Now that you know which county has the highest population, is that city with the highest population in that county?

• What city in the contiguous United States has the highest population?

1.   ***Click*** the **Clear Selected Features** button ⊠ on the **Tools** toolbar.

2.   ***Turn on*** the **Cities** layer in the table of contents by checking the box beside ☑ Cities .

3.   Right click the ☑ Cities layer and select Open Attribute Table.

4.   ***Sort*** the table by population by ***right clicking*** on POP1990 field header box and ***selecting*** ⫣ Sort Descending .

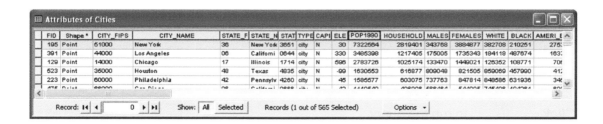

The answer lies on the opposite coast!

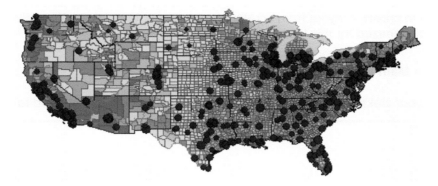

5.   *Click* the **Clear Selected Features** button on the **Tools** toolbar.

Selecting More Than One Feature at a Time in ArcMap

  • Which state(s) have the fewest number of people per square mile?

Let's switch gears and try to identify places that are much less populated. You will conduct a search that will show which state or states have the lowest population density.

1.   *Right click* on the ☑ States layer in the table of contents and **Open Attribute Table**.

2.   *Scroll* to the **right** and *find* the **field header** POP90_SQMI.

3.   *Right click* on that **field header** and *sort* **ascending** to find the smallest number.

There are two states that both have 5 people on average per square mile!

4.   *Select* the first state by clicking on the **row header box** on the far left side of the table. This will highlight the row and that state on the map.

5.   To select the second row also, *hold* the **Ctrl key** and *left click* on the second **row header box**. This will highlight the second row and a second state.

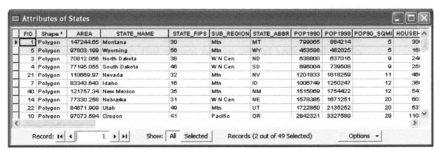

6.   *Click* to **minimize** the table to see the map with **Montana** and **Wyoming** highlighted.

7.   Click to save the ArcMap document.

Making a Layout Using ArcMap

In order to communicate your results to others, one of the best ways is to produce a map layout. In addition to showing your map, this map layout will need to include all important map details such as a title, legend, directional arrow, scale, and author. With a program like ArcMap, you have the option to create a layout from scratch or use preset templates. You will use a template that has been customized for this lesson.

1.   From the **Main Menu**, *click* on **View**, and then *select* 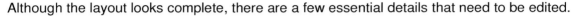 .

Although the layout looks complete, there are a few essential details that need to be edited.

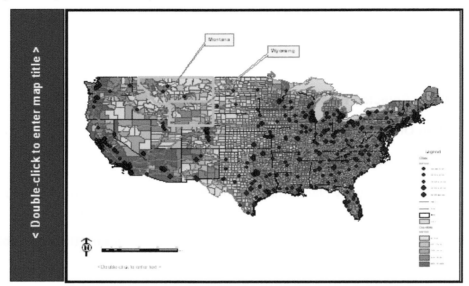

2.   *Double click* on the text that reads: **<Double-click to enter map title>**.

3.   When the **Properties** dialog box appears, *type*
     States with Lowest Population Density.

4.   *Click* ____OK____ .

5.   *Double click* the text that reads: **<Double
     click to enter text>** in the lower left corner of
     the map display.

6.   *Enter* your **name** and **today's date** in the
     Properties dialog box.

7.   *Click* ____OK____ .

You are now ready to print the document.

Printing a Map using ArcMap

The size of the map that you print will be determined by the purpose of the map that you need for your application or presentation. In this introductory lesson, you simply need to be able to print a map on a standard piece of printer paper.

1.	From the **Main Menu**, *select* **File**, **Print Preview**.

2.	*Verify* that all parts of the map are included within the border of the paper.

3.	*Click* Print... to **print** the document.

The final print out should look like this:

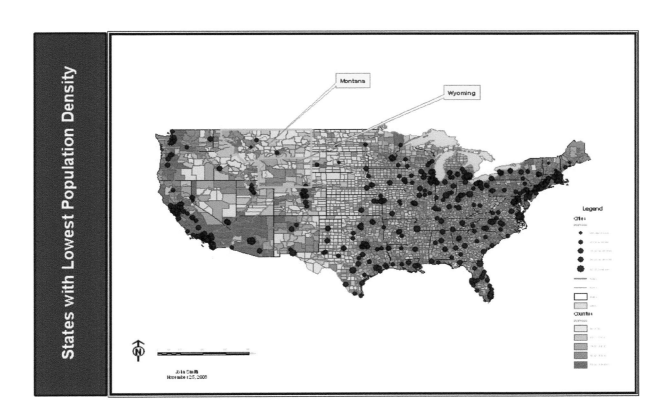

Lesson 2:
An Introduction to the Global Positioning System

The **Global Positioning System (GPS)** is a worldwide radio-navigation system formed from a constellation of satellites and their ground stations. Although initially designed for US military for defense purposes, it is now available for the general public to use to determine a specific location and to navigate to it on land, water, or in the air. It can also be used to record locations of land features on the Earth's surface for mapping features.

Common applications for GPS technology include navigation and identifying location. These applications GPS Technology allow users to obtain immediate and continuously updated information about their location. Have you ever tried to find a house address in the dark? Finding an address during the day can be challenging if you are unfamiliar with an area. After the sun has set, and the street signs and house numbers are not as easy to read, it can be extremely difficult. Newer navigational devices can lead you to an address – some with audio directions – by giving you step by step directions to your destination. Even GPS units can be programmed to display directions to a destination either by address or coordinates.

Common Navigational Terminology

Navigational devices have been around for many years. Long before GPS units, people were navigating the world using less technologically advanced devices such as simple compasses and maps. It is nice, however, to have a device which has the capability to contain a compass, a map, an altimeter, odometer, speedometer and more! Even though GPS Technology is an advancement in navigation, it is based on principles of earlier technology, and there are some typical navigational terms that are the same whether you are using a plastic card compass or a high dollar GPS model when it comes to navigating.

When using any type of navigational equipment, there are some common terms that are helpful to know to determine where you are and where you are going. The direction traveled from your starting position to your destination is called a **course**. Whether you are traveling north, south, east, or west, the direction you are moving in is called **heading**. Have you ever heard the expression, "I need to get my bearings"? If you were traveling along a course and stopped to shop or take a photo, then you have to travel now from that position to your destination. **Bearing** is the term for the direction traveled from your current position to your destination. Knowing a path's **elevation**, or altitude above sea level, can be

essential when you are hiking in an area. Knowing these terms can allow you to properly communicate to others using similar navigational equipment.

Common GPS Terminology and Facts
In recent years, Global Positioning System units have become more popular and affordable. Although their uses can vary from recreational to professional, there are some common elements and terminology that apply to all GPS units. GPS units are able to obtain real time location data making navigation easier. Some GPS units are able to record data with the push of a button saving points of interest for future reference. **Waypoints** are locations or positions of landmarks or other points of interest that can be stored on a GPS unit. The GPS user can actually go to that point and record the waypoint by using the GPS unit, or a coordinate or address can be plugged into the GPS unit. A **route** is a group of waypoints in their navigational order that can be entered into a GPS unit. A **track** is the current direction traveled relative to a ground position.

Did you know that GPS units currently receive signals from over 40 satellites that are included in the GPS

constellation? If your unit has a satellite page, you can see exactly which satellites, by number, your unit is receiving the signals from and the strength of the signal being received from that satellite. The satellite numbers will change as you move and as they continue on their orbits.

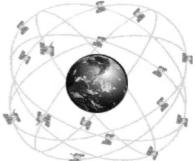

Many GPS units also have an interface that is very similar

to a compass. It displays a face that appears as a compass, showing the cardinal directions, north, south, east, and west with rotating directional arrow. The differences between the old

www.gsfc.nasa.gov

a

compasses and the GPS is that the GPS can also let you know at what speed you are traveling, possibly your current altitude, and constantly update the direction in which you are traveling.

Using the eTrex Legend
In this portion of the lesson, you will become familiar with the eTrex Legend. Take a moment to familiarize yourself with the eTrex Legend GPS unit.

1.   ***Turn*** the unit on by ***holding*** the **Power** button down for three seconds.

       The unit will first search for satellites. A message will appear that says "Poor Satellite Reception".

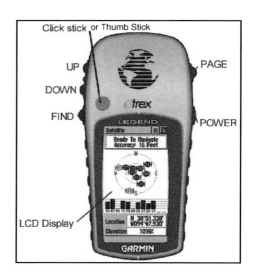

2.   Using the **Page** button, look at the five main screens that the eTrex Legend has:

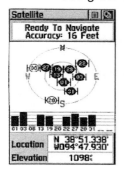
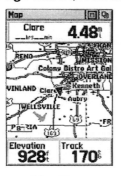

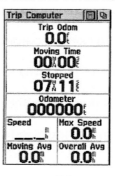
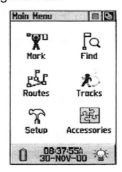

Satellite Page   Map Page   Navigation Page   Trip   Main Menu

You will need to verify that the unit is set to the correct time zone.

3.   From the **Main Menu** screen, *select* Setup.

4.   Using the Thumb Stick, select Time.

5.   If it is correct, press the Page button twice to go back to the Satellite screen.

6.   If it is incorrect, use the Thumb Stick once more to navigate to the Time Zone.

7.   Depress the center of the Thumb Stick to display the available Time Zone options.

8.   Select the correct Time Zone, then press the Page button twice to go back to the Satellite screen.

GPS Activity - Using a GPS unit

Each group of 2-4 students will use a Garmin eTrex Legend GPS unit. Each group will also need a copy of the S1U1L2 GPS Activity Worksheet to record answers on.

In order to get the best satellite signal possible, you should have a clear view of the sky and hold the unit flat and away from your body. Use the following instructions to use the GPS unit:

1.   ***Turn on*** the unit by pressing the **POWER** button on the side of the unit.

2.   When the unit is initially turned on, the **Satellite Page** appears as the unit begins receiving signals from satellites within its range. The **Satellite Page** provides information on satellite tracking, informs you when the unit is ready and shows your location coordinates. There is a number assigned to each of the satellites in orbit. The satellites within the reception range of your GPS unit show up on the screen in their positions in the sky. The center point of the circle represents the point directly above your head. As you move out away from the center to the outside rings, the satellites are positioned towards the horizon. The satellites sitting on the outside circle are orbiting on the horizon, and your GPS unit may have difficulty receiving a strong signal from them. The bars at the bottom of the screen show the strength of the signal received from each satellite. The GPS unit will not display a longitude/latitude position until it has received a strong enough signal from at least **4** satellites. You will get a message on the screen once a position is determined.

3. Select a location (away from the building and other groups).

4. Record the first set of coordinates on the S1U1L2 GPS Activity Worksheet.

5. Move at least 15 feet away to another location and fill in the chart for Waypoint #2.

6. Continue this process for the last two waypoints.

7. Answer the rest of the questions on the Activity Worksheet (front and back).

Lesson 2 GPS Activity Worksheet

As you complete the activities for the GPS lesson, enter your information in the appropriate spaces:

		Coordinates	Elevation	Accuracy
Waypoint 1	N			
	W			
Waypoint 2	N			
	W			
Waypoint 3	N			
	W			
Waypoint 4	N			
	W			

Sketch a map of the location used for area measurement and place stars with the waypoint number in the locations where you stopped to take your readings.

Unit 1, Lesson 2

As a group, answer the following questions.

1.   Do you think the readings were accurate? Why or Why not?

2.   How can being too close to a building affect your GPS unit?

3.   Do you think it matters if you hold the GPS unit an arms length in front of you versus an arms lengthy out to your side? Why or why not?

4.   Can you think of at least 5 professions that might use a GPS. Briefly explain how that profession uses it.

Lesson 3:
Introduction to Remote Sensing

What is Remote Sensing?

Remote sensing, simply put, is observing something from a distance. In geospatial studies, it can mean using air photos and satellite images as a means of studying an area. By using imagery of the Earth, areas of the world can be studied and assumptions can be made about land cover without actually traveling to a location. This is important if the area you are observing is too treacherous to physically go to or just too far away.

Landsat Satellite image of Essex County, NJ

Interpreting what you see in an image can be difficult initially. For instance, how many baseball fields do you see in the image on the right? There are at least two; one on each side of the river. How many bridges do you see? At first glance you may say three but there are actually four. There are many techniques that can be used to modify images to enhance certain parts of the image for interpretation; whether it be to better show healthy vegetation or to examine the extent of damage in a burned area. True color images, like the one on this page, are similar to what our eyes see. For this lesson, you will study an area using this format of photography.

Heads-up digitizing

For just about every square foot on this earth, you can find information. Data of all different types can be found about a place just by browsing around on the Internet or in a public library. Think about where you are right now. Did anything historically significant happen there? Did anything historically significant happen in your community? Do you know what the temperature is right now? What is the coldest it has ever been in that area? The warmest? Answers to questions like these as well as many others can be represented in digital format on a map using geospatial technology. Not all information however is in a format that can be automatically displayed on a map. What happens when you find an area that does not have geospatial information available? You can either study another area or create the data yourself. Creating this data might involve different methods such as using a GPS unit to collect information or another way which uses remote sensing, heads-up digitizing.

Heads-up Digitizing and GPS

One way to collect your own geospatial data is to use high quality GPS (Global Positioning System) units to make detailed point, line and polygon data sets. To do this, a person must go to a site and obtain these readings. But what if you need to collect data in an area that may be considered hazardous? If the terrain is not safe due to an area that has possibly been washed out or if a chemical spill has occurred in a region, other means of documenting an area may need to be considered. Another consideration may be time. Do you have time to travel to the destination in question? It might take two days to get to a destination and then two days back again. A final consideration might be cost. Traveling can be expensive especially when you have to factor in additional costs like food and lodging also. If you can get the same result without having to spend money on travel, equipment, and sending trained personnel to do a job, it might be worth it to stay home and use heads-up digitizing.

With heads-up digitizing, you use an air photo or satellite image and trace or outline features on an image to create shapefiles. Although the process might sound easy to do – and for the most part is – it does require that the user be consistent in their interpretations and is very detailed oriented in what they capture. In this lesson, you will use heads-up digitizing to update a map for a college campus; thus introducing you to one of the many applications of remote sensing.

Working with Heads-up Digitizing - Scenario

Two months ago, a Category 2 hurricane moved through La Playa, a small resort and college community on the Gulf of Mexico. The residents of this area consider themselves "lucky" with the majority of the area only suffering from wind damage. Although numerous roofs will have to be repaired or replaced, water damage was at a minimum and there was no tidal surge to report - this time. La Playa Community College is considering this a wake-up call for their administration. They have proper insurance on all of their facilities but do not have proper documentation on each of the buildings with true map scale. One of the faculty members, who recently moved, began a

Blue tarps draped over damaged rooftops in La Playa.

digital map a couple of years ago but did not have time to complete it before moving to another state. To make matters worse, a Category 5 hurricane (with winds exceeding 156 mph and a tidal surge exceeding 18 ft) is predicted to make landfall within the next week and La Playa lies in the center of the cone of uncertainty. The administration is calling for an immediate updated production of a digital map of the campus. You have been asked to update this map but due to the impending weather situation, do not have time to travel to the campus to take GPS readings. A satellite image has been supplied to you for this endeavor.

Making a Digital Map of the Campus

To update this map of the campus, you will use a satellite image in ArcMap. In summary, what you are doing is using remote sensing to modify shapefiles and then create a map layout of the campus. NOTE: Data from this exercise is from genuine locations with the names fabricated. The name La Playa is fictional.

1.   *Start* ArcMap by *double-clicking* the [ArcMap] shortcut on your computer desktop (or by *clicking* the Windows Start button **start**, *pointing* to **All Programs** ▷ ArcGIS ▶ and *selecting* ArcMap .)

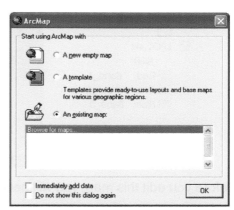

2.   When the ArcMap window appears, make sure that the radio button next to the (Start using ArcMap with) ⊙ An existing map: option is selected.

At the bottom of the ArcMap startup dialog box, make sure that Browse for maps... is highlighted.

3.   *Click* OK .

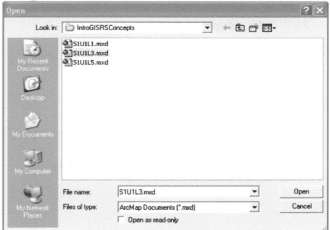

4.   In the **Open** dialog box, *navigate* to the **C:\STARS\IntrotoGISRSConcepts** folder.

5.   *Select* the **S1U1L3.mxd** file.

6.   *Click* Open .

A **map document** will open displaying an image of La Playa Community College. On the left side of the ArcMap window is a pane with the map layers listed. This is known as the Table of Contents. The first three layers that are listed are not displayed at this time in the map window because they are turned off (or not checked). These show the motor vehicle entrances and exits, streets, and campus buildings. The fourth layer, which is selected, is the image that you see in the ArcMap window. This displays the entire LPCC campus.

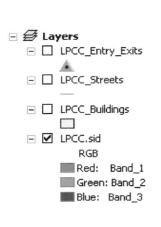

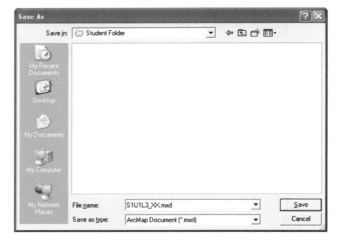

Before you edit this map, you will need to save it into your student folder.

7. To **save** this ArcMap map document in your student folder, **select** Save As... from the **File** menu.

8. **Navigate** to a location specified for you to save your work on your computer hard drive. **Name** the new file **S1U1L3_XX** (where **XX** is your initials) and **click**
   Save.

Editing Shapefiles in ArcMap

Each of the shapefiles listed in the table of contents will need to be edited so that they represent the most current layout of the campus.

1. **Turn on** the **LPCC_Entry_Exits** layer by checking the box to its left in the Table of Contents.

This will show the points of entrances and exits on the campus for motor vehicles. This is important information to have on hand should a quick evacuation from the campus ever be deemed necessary. Though there are three points documented, several more need to be added to this file. In order to edit, you will need to open the Editor toolbar.

2. From the **Main Menu**, *select* **View** then **Toolbars**.

3. From **Toolbars** drop down menu, *select* **Editor**.

4. *Select* Editor ▼ on the **Editor** toolbar to expose the drop down menu.

5. *Click* Start Editing .

6. *Verify* that **Task** is **Create New Feature**.

7. *Verify* that **Target** is **LPCC_Entry_Exits**. If either of these is different, click the down arrow to the right of these boxes and select the correct option from the list provided.

Your Editor toolbar should look like this:

8. *Click* the **Sketch Tool** button on the **Editor** toolbar.

9. *Single click* on the map in a location that designates the entrance or exit for a motor vehicle.

    The first point will appear with a blue dot with a red X on it.

10.   **Single click** on another spot on the map that designates an entrance or exit for motor vehicles. **Continue** this process until you have points at all of these places.

11.   To complete this process, **click** Editor ▼ to display the drop down menu.

12.   **Select** Stop Editing from the list provided.

A Save box will appear.

13.   **Click** Yes to save your edits.

14.   **Click Save** 💾 to resave your ArcMap document.

To erase a point....
• If the point is still highlighted like the one above, simply strike Delete on your keyboard.
• If it is any other point, click on the Edit tool ▶ and then click on the point you wish to remove. Once it is selected, strike Delete on your keyboard to remove it.

Your map will look like this:

The second shapefile you will need to edit is the Streets shapefile.

10.   **Turn on** the **LPCC_Streets** layer by checking the box to its left in the Table of Contents.   ☑ LPCC_Streets

The street file will open, however the streets directly north of the campus and to the east of the campus are not included in this file. Both will need to be added.

11.   **Select** Editor ▼ on the **Editor** toolbar.

12.   **Click** 📝 Start Editing to begin an editing session.

13. *Change* the **Target** to **LPCC_Streets** on the Editor toolbar.

Start on the street to the east of the campus along the treeline.

14. *Single click* on the point at the north end of this road.

15. *Drag* the mouse down until you meet the street at the bottom of the screen running east to west.

16. *Double click* to finish the line segment.

    If you do not like your work, simply strike Delete on your keyboard and try again.

    To view the western end of the road at the top of the map display, you will need to zoom out.

17. On the **Tools** toolbar, *click* the **Fixed Zoom Out** tool ⤢ **twice**.

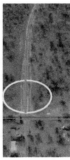

18. Starting on the western end of the road, *single click* at the point where the street will begin.

19. *Drag* your mouse up until you meet the first curve.

    Curves require a little more concentration and precision. You will need to click as many times as you need to make the street follow the curves.

20. *Clicking* numerous times, follow the curves in the street.

21. End this street by *double clicking* at the beginning of the next street segment.

Although your new street will show only as highlighted, this example is exposing all of the different clicks it can take to make the street "curve" to follow the street in the image.

22. To save your work, *click* and *select* **Stop Editing** from the drop down menu.

23. *Click* ⎣___Yes___⎦ to save your edits.

One last shapefile that will need editing will be the campus buildings.

24. *Turn on* the **LPCC_Buildings** layer by checking the box to its left in the Table of Contents.

☑ LPCC_Buildings
☐

Many of the buildings on the campus have been digitized, however those in the northwestern portion need to be added. Each of the buildings has a shadow on the northern side of it. Be careful not to include the shadows in the polygons that you will draw for these.

25. *Select* **Start Editing** from the drop down menu on the **Editor** toolbar.

26. *Change* the **Target** on the **Editor** toolbar to **LPCC_Buildings**.

It will be easier to distinguish the sides of the buildings if you zoom in closer.

27. *Click* the **Fixed Zoom In** tool until the buildings that need to be added to the map are closer.

28. *Use* the **Pan** tool to *click* on the image and *drag* the rectangular building to the center of your ArcMap window.

29. On the **Editor** toolbar, *click* on the **Sketch** tool .

30. *Single click* on one of the four corners to begin.

31. *Drag* your **mouse** to another corner and *single click* again.

32. *Drag* your **mouse** to a third corner and *single click* and then to a fourth corner and *single click* again.

33. *Drag* your mouse back to the original corner and *double click* to create the building polygon.

34. **Follow** these procedures to **create polygons** for the remaining buildings in this area. Some buildings are going to require using more than four straight lines. For these you will have to click multiple times as you did when creating curved lines for the streets.

35. **Select Stop Editing** from the **Editor** toolbar.

36. **Click** ___Yes___ to **save** your edits.

37. **Click Save** 💾 to **resave** your ArcMap document.

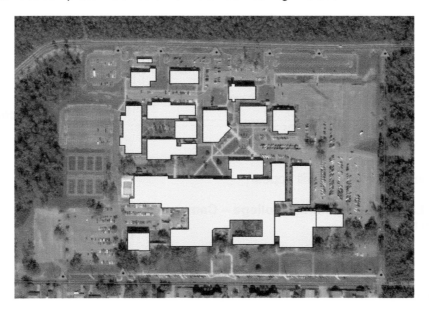

38. To return back to the original extent, **select Bookmarks** from the **Main Menu**.

39. **Select Campus** from the drop down list.

Your map is almost complete! It should now look something like this:

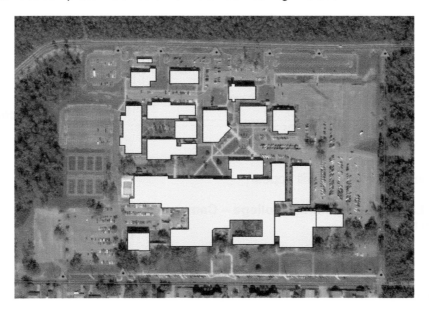

The final step of the editing process will require the completion of a map layout. This layout will enable you to display the shapefiles that you just edited. Because you need just the map showing, you will no longer need the image in the background.

40. **Uncheck** the box beside LPCC.sid to turn off that layer.

☐ LPCC.sid
RGB
■ Red: Band_1
■ Green: Band_2
■ Blue: Band_3

Editing a Layout in ArcMap

1.   *Switch* over to **Layout View** by *clicking* on the **Layout** button ⬚ at the bottom of the ArcMap window.

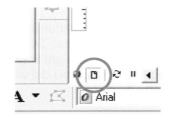

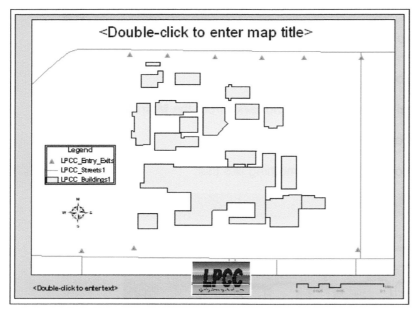

A layout will appear but it is not complete. You will have to edit the title, add your name and today's date, and change the map scale to a different model.

<Double-click to enter map title> .

A Properties dialog box will appear.

2.   *Double click*

3.   *Enter* **La Playa Community College – Campus Map** in the Text box.

4.   *Click* ⬚ OK ⬚ to *apply* the change.

5.   *Double click* <Double-click to enter text> at the bottom of the map layout.

6.   *Enter* **Your Name** and **today's date** in the Properties dialog box.

Last, you are going to change the scale bar to a different model. The current scale is not wrong, it has been suggested that a different model would be easier for end users to interpret.

---

7. **Double click** on the current scale bar at the bottom of the layout.

The Stepped Scale Line Properties box will appear.

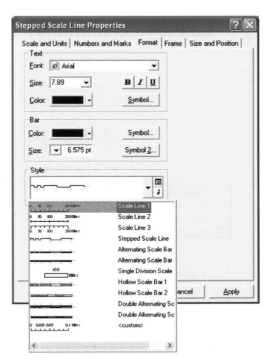

8. **Click** the **Format** tab at the top of the screen.

9. **Click** the down arrow under Style to expose a list of options.

10. **Select Scale Line 1** from the list by **single clicking** on it.

11. **Click** OK to **apply** the change.

Your map is now complete.

Printing a Map using ArcMap

The size of the map that you print will be determined by the purpose of the map that you need for your application or presentation. In this introductory lesson, you simply need to be able to print a map on a standard piece of printer paper.

1. From the **Main Menu**, **select** **File**, **Print Preview**.

2. **Verify** that all parts of the map are included within the border of the paper.

3. **Click** Print... to **print** the document.

The final print out should appear similar to the map below:

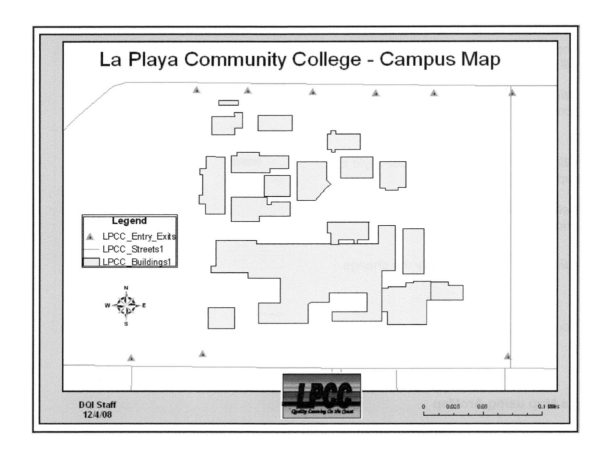

La Playa Community College - Campus Map

Legend
▲ LPCC_Entry_Exits
— LPCC_Streets1
☐ LPCC_Buildings1

N
W ✦ E
S

DOI Staff
12/4/08

LPCC
Quality Learning On the Coast

0   0.025   0.05   0.1 Miles

Lesson 4:
Introduction to Surveying Technology

All applications of geospatial technology benefit from precise instruments and tools to provide the best analysis and decision making. If a rescue helicopter is locating a damaged ship or a hiker is mapping trailheads or a soil scientist is mapping soil types across an area, these individuals can accurately map and find these locations within 3 to 5 meters of their actual location using the GIS, Remote Sensing, and GPS. In each of the above situations, locations or events must be found quickly and easily.

Sometimes a higher degree of accuracy is required over the need for ease and speed. The Chunnel - connecting France and England via a tunnel under the English Channel – was completed by using two drills beginning at opposite ends of the proposed tunnel. It is essential that the two teams meet at the same spot to complete the tunnel where three to 5 meters can be the distance that defines success or failure. This degree of accuracy can be found through surveying technology.

Surveying is a component of geospatial technology that allows for even greater accuracy (within millimeters) by establishing locations and points of reference for accurate mapping. Fundamentally, surveying calls upon geometry, trigonometry, and other higher mathematics to find precise distances and angles between points on the Earth. When combined with GPS technology, surveyors can attain this accuracy faster and while in motion, provided with a clear view of the sky.

Surveyors are valuable in any situation where near exact measurements are required or a large area needs mapping for a geospatial project. These projects include precision measurements for transportation and infrastructure development. Similar to the Chunnel example, precise measurements are needed to plan a road, canal, or rail. Surveyors are used to map wide areas like the ocean floor using advanced surveying/GPS technology or map the land as the USGS and even Lewis and Clark have. Our legal system calls upon surveyors to settle civil disputes that require accurate boundary measurements. These disputes can be simple clarification of whose property a fence may lie on or determining larger disputes over taxes when it is unclear which city county or state a business is established.

Individuals in this discipline use a variety of tools. Common tools may be simple tools like a measuring tape and compass or more advanced tools like the theodolite, which measures angles. The more advanced tools include the Total Station, which has the capabilities of a theodolite and a GPS receiver. For this lesson, we will use a measuring tape to create an accurate map of an enclosed space.

Here is your first opportunity to create a map yourself! You should choose an interior space that contains at least two features within the room to map in addition to the boundary. Follow the directions below to create a map of your own.

Materials Required

Graph Paper: Graph paper is ruled with horizontal and vertical lines to form a grid making it ideal for drawing points, lines, and polygons that will represent features in your space. For measurement purposes, graph paper with four lines per inch forming ¼" squares is suggested, since it is an English measurement that is found on most rulers to which you have access. There are other types of graph paper in other measurements like centimeter sized squares, but this may be more difficult to manage. You will likely be measuring your space in feet and converting to centimeters on paper may be more complicated than necessary. Like wise, if you measure your space using a metric instrument, then centimeters may be ideal.

Measuring Instrument: Any measuring instrument that will allow for measurements of the length of the longest wall is suitable. Tape Measures with divisions of feet, inches, and sub-inches are likely the most readily available.

Colored Pens or Pencils: Your space may be filled with many types of features that can best be illustrated with a unique color for each feature. Features may include desks, computers, chairs, cabinets, tables, or any other object taking up space in the room. A multi-colored map is preferable, but you may also utilize shading, cross-hatching, or other patterns to distinguish features.

Assessing the Area for Measurement

You will select an interior area with at least two features to survey and map. The room should be a simple shape (or polygon) with four sides, which will be the boundaries for your map. Selecting a room with curved walls, with many columns, or other detail will increase the difficulty of the survey. For this exercise, you should also select relatively simple features within the room like tables or desks. If you select a feature like personal computers, you may define a general area for the entire feature. For example, a computer may consist of a monitor, tower, mouse, keyboard, and other peripherals that can be generally defined as: an entire workstation is 3 feet by 3 feet. You are the surveyor in this lesson, so you must define the detail with which features are mapped.

Taking Measurements

This lesson assumes that graph paper with ¼" squares is being used with an English Tape measure in feet and inches down to a 1/16". The space in the example will be a rectangular room with several tables, chairs, and desks.

1.   Begin by taking measurements for the walls (the boundary of your space) in feet from corner to corner on the floor. Measuring on the floor will ensure that the tape measure does not bend and remains taught ensuring accuracy. Number each side and record them to the corresponding name on the "Survey Recording Log" included at the end of this lesson.

Surveying is about precision. Measure all four walls rather than measuring two and assuming a true rectangle. What do you expect to find? It is not uncommon that two parallel walls thought to be of equal length are found to be slightly different lengths. After measuring all four walls, did you find 2 equal sets of measurements or four different measurements?

2. Record your findings on the Survey Measurements Recording Log.

3. Select the first feature to include in the map. This example will use a desk.

4. Measure the dimensions of the top surface of the desk

   Remember a map is drawn from the orthogonal viewpoint (from above). This example assumes that the top of the desk is the only thing visible from above.

5. Record the length and width in the Survey Measurement Recording Log

6. Select a side along the length of the desk and measure from that side to the nearest wall side along the land also a side along the width of the desk measure from the corner to the nearest wall. Record these measurements in the Survey Measurement Recording Log.

   This measurement assumes that the desk is parallel to all four walls. For rooms where objects are not parallel, you must know the distance from the long and short walls for at least two points.

7. Follow the same procedure for the remaining objects in the room. If you choose to include a computer workstation in your survey, it is acceptable to estimate the average size of the computer an its peripherals. For example, a desktop tower, monitor, mouse, and keyboard may be estimated to a size of 3' x 3'.

Scaling Your Measurements

1. Using a 1 foot = ¼ inch scale[1], draw the boundaries (walls) of the classroom on your grid paper. *Hint: Before you begin drawing on your grid paper, make sure that you count out the grid cells that will be needed to layout the boundaries first. You may find that you need to start closer to the edge of the paper to fit the whole room on one page.*

2. Using the same scale draw each feature by the appropriate size and position with respect to side 1 and side 2.

3. Shade each or color each feature or feature type, so that it is clear which feature is represented.

3. After you have finished mapping the objects in your classroom, create a legend for your map to specify which color represents which object. Also add a visual scale by using a bar scale to represent it on your map.

4. Turn in the map, with legend and scale, to your instructor. Get more grid paper from your instructor to practice mapping a room in your home.

---

[1] *Each one foot square on the ground is equal to one ¼" square on your grid paper.*

The finished map should look similar to the map below:

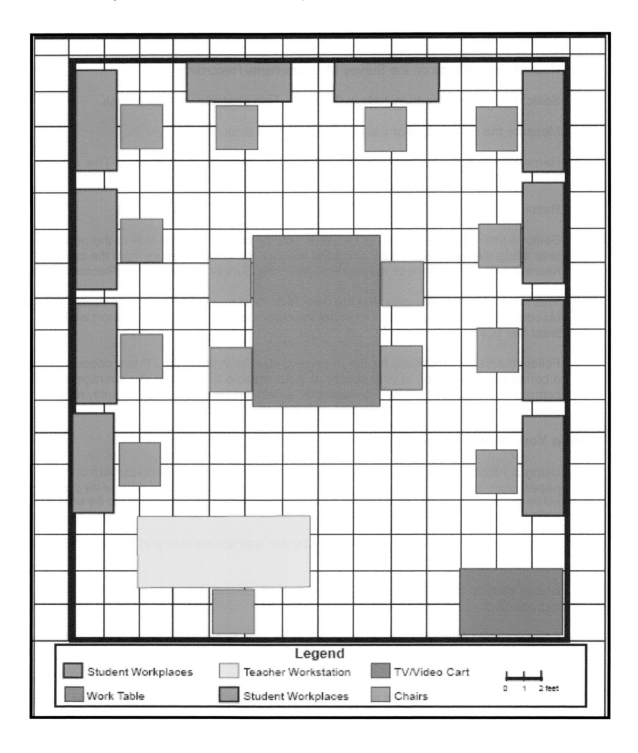

Legend

Student Workplaces	Teacher Workstation	TV/Video Cart
Work Table	Student Workplaces	Chairs

0   1   2 feet

Survey Recording Log

Student Name(s):_____

Boundary	Length	Scaled Length (1 Ft = ¼" or 1" = 1/48") Assumes Graph Paper with ¼" Squares		
Sample 1	28 ft.	28 Squares or 7"		
Side1				
Side 2				
Side 3				
Side 4				
Feature Name	Length	Width	Distance from Side 1	Distance from Side 2
Example: Table	*9'*	*3'*	*5'*	*6'*

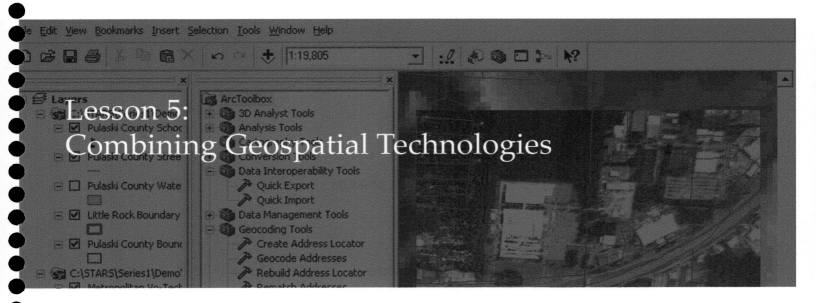

Geospatial Technology has been shown in this unit as four major disciplines that can be utilized to make decisions and enhance our understanding of the Earth and its resources and features. A successful analysis calls upon all of these disciplines and combines them to provide complete results. This lesson will give you an application of how that analysis could benefit Law Enforcement agencies.

To track recent residential burglaries in the Burlington County, NJ area, the police chief has requested a map showing where the bulk of the activity affecting residential areas has occurred. The goal is to increase patrols in the area during the peak times that the offenses have occurred. A file with addresses for these crimes is available, but it is not clear if they occurred in residential areas. Land parcel data is available for this area showing the data for three types of land use for this county; residential, commercial, and utility.

With the aid of parcel data from two surveying companies, you will be able to view the entire community. The data from both however will need to be combined into one map of the area.

Combining Survey Technology with GIS
To begin this lesson, you will need to start with a basic map of the study area. There are two main survey companies in this area and both have supplied their land parcel surveys for this study. Thankfully, both companies use the same geographic reference system – in this case, State Plane - to build their numerous sets of data in. These surveys will need to be combined into one file so that further analysis will be possible.

Gathering basic data and combining it to make a base map of an area is one of the first steps in project implementation. The task of taking data from more than one source, in possibly multiple projections, and combining it into one workable data set is a common task for those who use geospatial technology.

Additional data will be supplied added as needed to complete the assigned task. This data will be supplied to you.

1. **Start** ArcMap by **double-clicking** the  shortcut on your computer desktop (or by **clicking** the Windows Start button ![start], **pointing** to **All Programs** ▶ 🗎 ArcGIS ▶ and **selecting** 🔍 ArcMap .)

2. When the ArcMap window appears, make sure that the radio button next to the (Start using ArcMap with) ⦿ An existing map: option is selected.

   At the bottom of the ArcMap startup dialog box, make sure that Browse for maps... is highlighted.

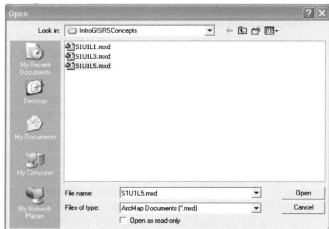

3. **Click** OK .

4. In the **Open** dialog box, **navigate** to the **C:\STARS\IntroGISRSConcepts** folder.

5. **Select** the **S1U1L5.mxd** file.

6. **Click** Open .

Merging Files in ArcMap

An ArcMap window will open displaying the land parcels in Burlington County, New Jersey. The data that you are viewing comes from two separate surveying companies, Brown and Smith Survey Companies. In order to search through these two sets of data, it will be much easier if you were dealing with one file. Merging these two together will complete this task. The Merge tool is located in ArcToolbox.

1.   ***Click***  on the **Standard** toolbar to open **ArcToolbox**.

An extra pane will appear on your ArcMap interface displaying the contents of ArcToolbox.

2.   ***Single click*** the **plus** sign to the left of Data Management Tools  to display the menu options.

3.   ***Single click*** the **plus** sign to the left of **General** to display several options including **Merge**.

4.   ***Double click*** 🔨 Merge to open the **Merge** dialog box.

5. In the **Input Datasets** box, *strike* the **down arrow** ▼ to display the two options available.

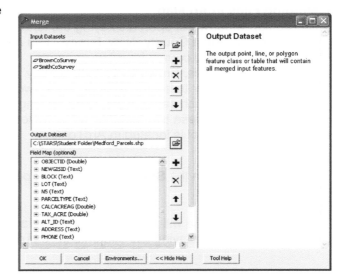

6. *Select* **BrownCoSurvey**.

7. *Strike* the down arrow once more and *select* **SmithCoSurvey**.

Both will now appear in the Input box.

8. *Single click* the folder 📁 to the right of the **Output Dataset** box.

9. *Navigate* to your **Student Folder** and *save* the file as **Medford_Parcels.**

10. *Click* [Save] to save this change and return back to the **Merge** dialog box.

11. *Leave* all other elements in the **Merge** box as they are.

12. *Click* [OK] to begin the **Merge** process. Once the process is complete, a Merge box will appear with "Completed" in it.

13. *Click* [Close] to finish the procedure.

14. At the top of the ArcToolbox window pane, *click* ✖ to close this small window pane.

This procedure created a new shapefile which will now have been added to your map and table of contents. This will be the topmost layer, covering the other two. Because you will not need the other two, you will simply need to remove them.

15. *Remove* the **BrownCoSurvey** layer by *right clicking* and *selecting*  from the options shown.

16. *Remove* the **SmithCoSurvey** layer by *right clicking* and *selecting*  from the options shown.

If an item is listed in the table of contents and is selected, when a map layout is made for printing or some type of distribution, all of these selected items will show in the map display and will also, by default, show in the legend. All items in the legend should be properly labeled so that anyone reading the map can understand what they represent. Although it is almost flawless, the Medford_Parcels shapefile, should be renamed with only a space in between each of the words.

17.   ***Open*** the **Layer Properties** for Medford_Parcels in the table of contents by ***double clicking*** ☑ Medford_Parcels .

18.   ***Select*** the **General** tab and under **Layer Name**, ***delete*** the underscore and ***replace*** with a blank space between **Medford** and **Parcels**.

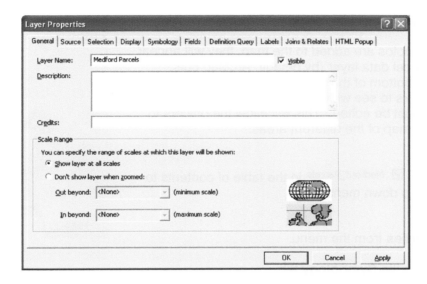

19.   ***Click*** OK to ***apply*** this change and ***close*** the **Layer Properties** box.

Using Remote Sensing Data to Enhance Survey Technology data in a GIS
There are six air photo files that are needed to display the entire area. You will add them to your current ArcMap document.

1. *Click* the **Add** data button ⬇ on the **Standard** toolbar.

2. *Navigate* to the **STARS/IntroGISRSConcepts** folder on your computer.

3. *Select* **medford_airphoto1, medford_airphoto2.sid, medford_airphoto3.sid, medford_airphoto4.sid, medford_airphoto5.sid, and medford_airphoto6.sid.** You may add them one at a time, or all at once by holding down the **Ctrl** button on your keyboard as you *single click* on each one.

4. After making your selection, *click* [Add].

   Once the air photos are added to the map, they will appear behind the parcel data layer (by default, ArcMap puts images at the bottom of the table of contents). In order to allow the officers to see which areas lie in which parcels, the parcel layer must be edited. This will allow the officers to see a realistic map of the different areas.

5. *Right click* on ☑ Medford_Parcels in the table of contents to display the drop down menu.

6. *Select* **Properties** from the menu.

7. *Single click* on the **Symbology** tab.

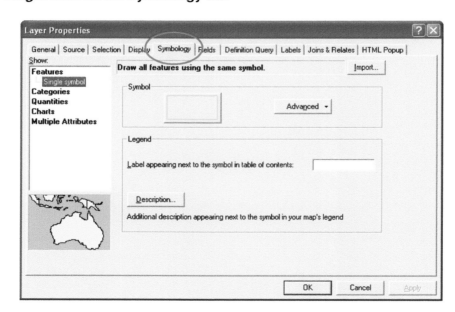

8.   ***Single click on*** the **Symbol** box.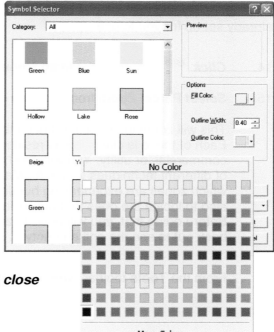

     This will display the **Symbol Selector** box.

9.   ***Single click*** on **Hollow** to change the appearance of the symbol.

10. ***Edit*** the **Outline Color** by ***clicking*** the down arrow beside it.

11. ***Select*** **Solar Yellow** from the color palette.

12. ***Click*** OK to close the **Symbol Selector** box.

13. ***Click*** OK to ***accept*** the change and ***close*** the **Layer Properties** box.

Your map display window will now appear like this:

To better view this land parcel data and the air photos below this layer, you will use a bookmark to zoom in. A spatial bookmark is a preset map extent that is saved in an ArcMap document. You can easily add and remove bookmarks as needed. In this map, there is a bookmark already set up for you to use.

14. *Click* Bookmarks from the **Main Menu**.

15. *Select* **Parcel Zoom** from the drop down list.

This will zoom in on the land parcels layer. Each of these is classified as residential, commercial, or utility. You will need a couple of different techniques to ask ArcMap to find which zones are experiencing burglaries and also to locate the ones that are residential. Before doing that, you will need to zoom back out to see the entire county.

16. From the **Tools** toolbar, *single click* the **Full Extent** ⬤ button.

Knowledge Knugget
Navigating in ArcMap

On the Tools Toolbar there are several buttons that have differing zooming and moving capabilities. Take a moment to explore the different tools. Return to Full Extent before continuing this lesson.

🔍➕	Zoom In	✋	Pan
🔍➖	Zoom Out	⬤	Full Extent
Fixed Zoom In		⬅	Previous Extent
Fixed Zoom Out		➡	Next Extent

Analyzing Point data in a GIS

Each call that comes in to the police dispatcher is documented in the police database. They are labeled as burglaries, robberies, or mva (motor vehicle accident). These are called Calls for Service because an actual person needs some type of first responder to come to their aid. To discover which parcels have the most burglaries, the calls for service database has been split up and the file that shows only burglaries is available for your research purposes. You will need to first add the Calls for Service file containing burglaries to your computer, find out which parcels these happened in, and then find out which of these are considered to be residential.

1. *Click* the **Add** data button on the **Standard** toolbar.

2. *Navigate* to the **STARS/IntroGISRS Concepts** folder on your computer.

3. *Select* **Medford_Burg** from the **Demo** folder.

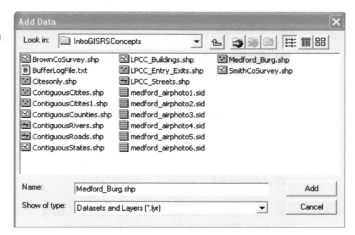

4. To *rename* this file, *right click* on **Medford_Burg** and *select* **Properties**.

5. *Click* the **General** tab and in the **Layer Name** box *rename* it **Medford Burglary Locations**.

6. *Click* **OK** to complete this process and *return* back to the ArcMap window.

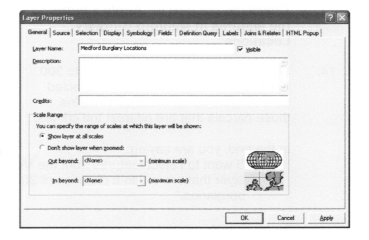

7.　To find out which parcels have had burglaries, *click* Selection from the **Main** menu.

8.　From the Selection menu, *select* ⧉ Select By Location... .

The **Select By Location** dialog box will appear. In this selection, you are building a statement that will allow ArcMap to search the data layers involved. This is also known as a query.

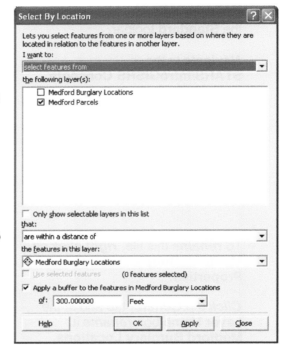

9.　In the **"I want to:"** box *confirm* that **select features from** is selected.

10.　In the **"the following layer(s):"** box *select* **Medford Parcels** by *checking* the box beside it.

11.　In the **"that:"** box, *select* **are within a distance of**.

12.　In the **"the features in this layer:"** box *select* **Medford Burglary Locations**.

13.　*Check* the box next to "**Apply a buffer to the features in Medford Burglary Locations**".

14.　*Change* the number of the **buffer** to **300 Feet**. Three hundred feet was selected as a suitable buffer distance to define those parcels that are nearest the crimes.

In the end, you are saying:
"I want to select features from the Medford Parcels layer that are within a distance of 300 feet of CFS burglaries."

9.　*Click* OK to allow ArcMap to complete this query.

These are the parcels that have had burglaries. Which of these are residential? You must now ask ArcMap to help you answer this question.

10. From the Selection menu on the Main Menu, *select* 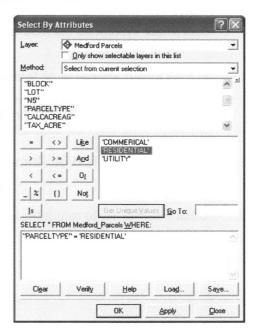 Select By Attributes... .

11. In the **Layer** box, *select* **Medford Parcels**.

12. In the **Method** box, *select* **Select from current selection**.

13. In the next box, *scroll* until you see "**PARCELTYPE**" and *double click* it. This will cause "**PARCELTYPE**" to appear in the bottom box if done properly.

14. *Single click* = which will also appear in the bottom box.

15. To display your options, *single click* Get Unique Values . The options for the different Parcel Types will appear above this button.

16. *Double click* on '**RESIDENTIAL**' thus making it appear in the bottom equation box.

17. Click OK to complete this process.

Your map will now display these residential land parcels that have had burglaries. The police force can see where the burglaries are taking place. The chief would like the addresses of these to be available so that he may determine which units can patrol these high crime parcels. In order to set up this report, more steps will need to be taken.

The first step will involve creating a shapefile using just the highlighted parcels.

17. *Right click* on **Medford Parcels**.

18. *Select* **Data**, **Export Data** from the list provided.

The Export Data dialog box will appear.

53

19.   *Confirm* that the **Export:** box contains **Selected Features**.

20.   *Leave* the '**Use the same coordinate system as:' radio** set to **this layer's source data**. This will allow the new shapefile that you are about to create to retain the same coordinate system as the Medford Parcels shapefile.

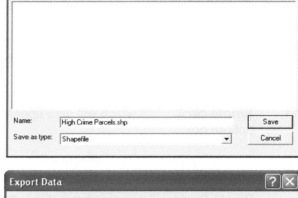

21.   In the **Output shapefile or feature class:** *click* and *navigate* to your **student folder**. This will tell ArcMap where to store this new shapefile.

22.   *Name* this shapefile **High Crime Parcels**.

23.   *Click* Save to return back to the **Export Data** dialog box.

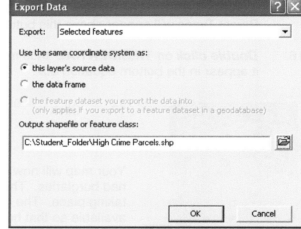

24.   *Click* OK to save the shapefile.

You will receive this box:

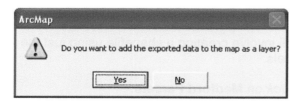

25.   *Click* Yes to add the High Crime Parcels shapefile to your current ArcMap document.

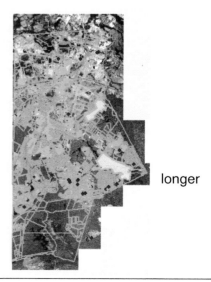

Because these are now their own shapefile, they no   longer need to be "selected" or highlighted.

26. **Single click** the **Clear Selected Features button** ⊠ on the Tools toolbar to clear them.

To make the High Risk Parcels easier to see, you will need to edit the symbology for them.

27. **Right click** on ☑ High_Crime_Parcels and **select Properties** from the menu provided.

28. **Select** the General tab (if not already selected).

29. **Rename** the layer **Medford High Risk Parcels**.

30. **Select** the Symbology tab next.

31. **Single click** on the Symbol box to display the Symbol Selector palette.

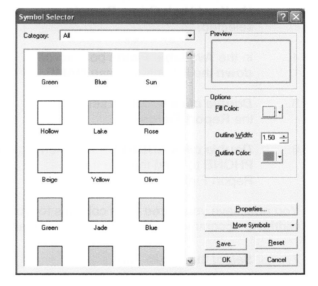

32. **Single click** on Hollow and change the **Outline Color** to Mars Red.

33. **Change** the **Outline Width** to **1.5**.

34. **Click** OK to return **Layer Properties** box.

35. **Click** OK to **close Layer Properties** and **return** back to ArcMap.

You map display appears similar to the one pictured here:

Now that you have a separate file for these parcels and it has been properly edited, you are now ready to set up a report for the chief of police.

Creating Reports in ArcMap
The report you are about to generate will supply the names and addresses for all residential land parcels that are within 300 feet of criminal activity. The chief wants all citizens that might be affected, to be alerted and made aware of the situation. He will also use this list to generate a more focused effort on patrol units in these areas.

1.　　From the **Main** menu, *select* **Tools**.

2.　　From **Tools**, *select* **Reports**, **Create Report**.

The Report Properties box will appear.

3.　　In **Layer/Table** *verify* that **Medford High Crime Parcels** is selected.

4.　　In the **Available Fields** box, *scroll down* the list until you see **NAME**.

5.　　*Double click* **NAME** to send it to the **Report Fields** box.

6.　　*Double click* **ADDRESS** and then **PHONE** to send them to the Report Fields box.

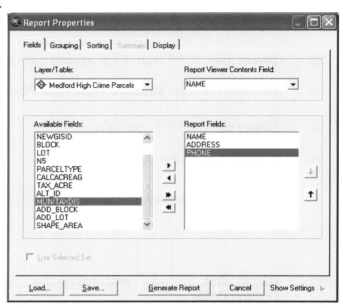

You now have your three main columns for your report. Next you will need to adjust the column widths to accommodate the information contained in them.

7.　　*Select* the Display tab.

8.　　*Click* on the plus sign in front of ⊞ Elements to display its menu.

9.　　*Check* the box next to ☑ Title. A list of current settings for a title will appear on the right side of the box.

10.　　Under the **Value** column, *click* in the box that says **Report Title**. The box will expand.

11.　　*Delete* **Report Title** and in its place *type* **Residential Areas Near Burglaries**.

12. ***Click*** the plus sign in front of ⊞ Fields   to display its menu.

13. ***Single click* NAME** to select it.

14. ***Change*** the **width** to **2** to allow 2" for the names of the residents.

15. ***Single click* ADDRESS**.

16. ***Change*** the **width** to **3** to allow 3" for the addresses of these residents.

The phone column will remain as is.

17. ***Click*** Generate Report to view the report.

The Report Viewer will appear showing your report. There are 18 pages of residents in these areas! What only took you less than an hour to produce would have taken much longer if someone had to look up all of these addresses individually. At this point you have produced the list and are ready to send it to the chief of police. Printing it out will not be necessary. You will simply need to save a copy in printable format to send to him.

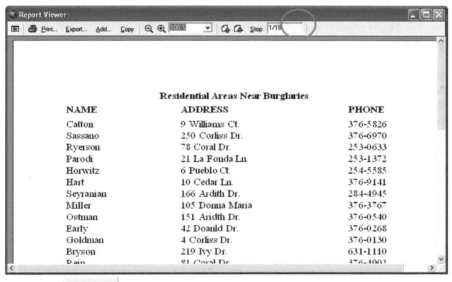

18. ***Single click*** Export... in the **Report Viewer** window.

19. ***Navigate*** to your **Student Folder** and ***export*** this report as **Residential Areas Near Burglaries**.

Not only does saving the report in Portable Document Format (pdf) allow you to be able to electronically send it to the chief, it also enables you to be able to add it to the ArcMap layout. Because of the size of this report, it is not advisable to add it to this map layout.

20.   ***Click*** the **Close** button ☒ to ***close*** the **Report Viewer** box.

21.   To ***save*** the report, ***click*** [Save...] .

22.   ***Naviagate*** once more to your **Student Folder** and ***save*** it as **Residential Areas Near Burglaries**.

Note: You will not see the pdf file that you recently saved in the Student Folder because the Save Report window will only show items that are in Report (*.rdf) format. At this time, there are none of those.

23.   ***Click*** [Close] to close the **Report Properties** box.

Managing a Layout in ArcMap
Creating the report was one means of communicating the answers to the questions posed to you. Another means to doing this is to create a map layout. This layout will give a visual guide to the officers to study that will compliment the information found in the report.

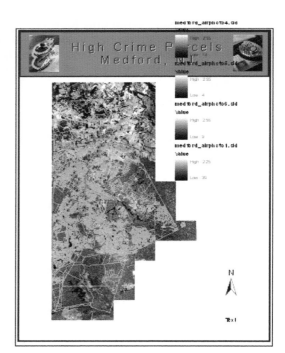

1.   From the **Main** menu, ***select*** **View**, then **Layout View**.

Your screen will switch to a layout that is partially been set up for you. There is a title, a legend, and a text box for your information. You will need to complete the layout by adding a scale and a compass and by editing the Text box and the legend before exporting it for printing.

2. ***Double click*** on the **Legend** to open the ***Legend Properties*** dialog box.

    It will not be necessary to have the air photos in the Legend. The following step will only remove them from appearing in the Legend, not the map.

3. ***Double click*** **Medford Burglary Locations**, **Medford High Risk Parcels**, and **Medford Parcels**. Each should now appear in the **Legend Items** box.

4. ***Leave*** all other boxes in the **Legend Properties** box as is.

5. ***Click*** OK to ***close*** the **Legend Properties** dialog box.

    Your Legend should appear like this:

    Legend
    ◇ Medford Burglary Locations
    ☐ Medford High Crime Parcels
    ☐ Medford Parcels

    You may need to move the Legend to a more desirable location.

6. ***Use*** the **Pan** tool  to **click** in the center of the **Legend** and ***drag*** it to a better location (where it is not overlapping the map or title).

7. ***Insert*** a scale bar by going to the **Main** menu, ***select*** Insert,
     .

8. In the **Scale Bar Selector**, ***select*** **Scale Line 1** from those listed.

9. ***Click*** OK to complete the selection.

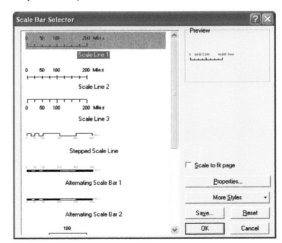

10. ***Click***, ***hold***, and ***drag***
     the **scale** to a position
     below the map.

One more missing or unfinished element to the map involves the author and date. Both are crucial information to those who may need to reference the map.

11. ***Double click*** on "**Text**" (seen at the bottom of the layout above).

12. When the **Properties** box appears, ***replace*** **Text** with **your name**, ***strike*** **enter** on your keyboard, then ***type*** **today's date**.

13. ***Click*** OK to ***close*** the **Properties** box.

14. ***Click*** 💾 to save your map.

Your map should now look like this:

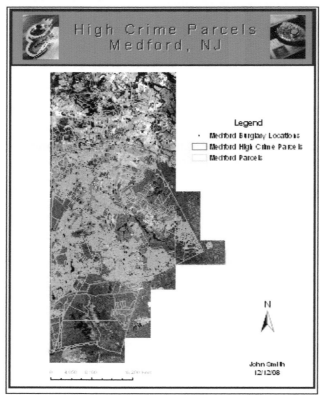

Exporting Files in ArcMap

The map files created in ArcMap, can only be viewed in ArcMap. In order to be able to electronically send a map to a person who does not own ArcMap, you must export the map layout as an image file. By doing this, a map layout can be inserted into a word processing program or into a presentation program. Exported layouts can be useful when added to a written report or an oral report to communicate project findings.

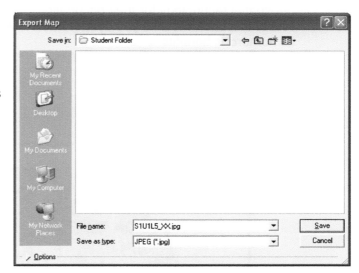

1. *Click* the ᴱile menu and *select* Export Map…. The **Export Map** dialog box will appear.

2. In the **Export Map** dialog box, *navigate* to the location on your computer that has been designated for saving your work.

3. If necessary, *select* the **JPEG (*.jpg)** format in the **Save as type** box.

4. *Name* the exported map **S1U1L5_XX** (where **XX** is your initials).

5. *Click* [Save].

Printing a Map Layout from ArcMap:

You will now print the map layout that you have created.

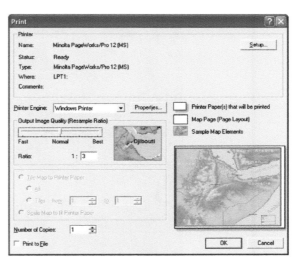

1. *Click* the ᴱile menu and *select* 🖨 Print…. The **Print** dialog box will appear.

2. *Verify* that the print settings are correct for your printer configuration.

3. *Click* [OK].

Saving & Closing:

1. *Save* 💾 the map document.

2. *Click* the ᴱile menu and *select* Eˣit OR *click* the **Close** ⊠ button in the upper right corner of the **ArcMap** window to exit.

You just completed a geospatial project using all of the components of geospatial technology to provide a service for law enforcement. Efforts like this save vast amounts of time, and therefore, resources that could be beneficial to another part of the law enforcement process. Gathering data (i.e. the Parcel data and the air photos) is one of the many steps involved in Project Management. Performing analyses and preparing the data for delivery as you did in this lesson are also parts of the Implementation phase and are vital skills for those who use geospatial technology for their occupations. Now that have been introduced to these skills, the rest of this book will take you deeper into the world of geospatial technology, its concepts, and its numerous benefits.

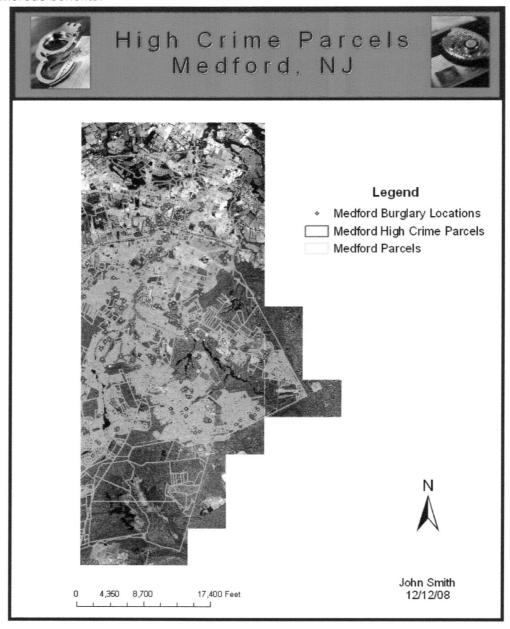

Introduction to Project Management

Introduction to
Geographic Information Systems
and Remote Sensing
Concepts

Learning Topics

Explore the Project Management
Model:
o Planning
o Implementation
o Presentation

Lesson 1:
Project Planning

The **Project Management Model** is the tool developed by Digital Quest to lead project managers through a set of steps, tasks, and techniques designed to help execute a thorough, successful geospatial project. The three phase process begins with **Project Planning**, flows into **Project Implementation**, and then concludes with **Project Presentation**. There is a defined flow in this model, and each part is critical for the entire model to be successful. This lesson will focus on the first component of the **Project Management Model** - **Project Planning**.

Project Planning involves successful completion of a list of tasks that will provide the foundation for the rest of the project. Each task is included to help ensure that any foreseeable obstacle that could pose a problem in the course of a project might be identified prior to **Project Implementation**. As you work through this list, keep in mind that you should be thorough in completing each one. The tasks associated with the **Project Planning** stage of project management include:

- **Problem Identification**
- **Project Stakeholders**
- **Project Objective**
- **Project Title**
- **Geographic Extent (or Area Of Interest)**
- **Project Feasibility**
- **Functional Requirements**
- **Project Design**

Project managers are presented questions and requests in various stages of development, but normally, regardless of the questions planning can be completed in sequence. Consider the following questions that you – as an employee of a GIS firm or Department - may get from a client or colleague:

"Should the city build a new high school?"

This could be a question posed to you by someone requiring GIS analysis. This question seems like a question asked in early stages of research. It offers several opportunities for research and is a problem that needs refining. (Remember that you have been asked for your perspective on the question.) This could lead to new questions: "Does the school system need to expand?", "Are current facilities sufficient to accommodate students?", or "Is a new school the

most cost effective solution?". When GIS is included in early stages of decision making, project managers can consider all possibilities in problem identification.

Now consider this question:

"Where should the city build a new high school?"

This question indicates that a project is further developed. The decision may have been made that the new school is necessary, and funds may have been allocated to begin the project. The question sounds more like a project objective rather than a problem that needs to be identified, but this question also requires focused research and problem identification. "How will the new school affect traffic patterns?" "Which school location will be most easily accessible to students?" "Are there any neighborhoods that could benefit more from having a school?"

As a project manager you should plan to use the full resources of the **project management model** for every question. To get a better understanding for each of these tasks, let's take a look at each one individually.

Problem Identification
Problem Identification is the act of defining the question to be considered, solved, or answered. This task is sometimes considered to be the most difficult one of the **Project Planning** process requiring extensive research before it can be established. Mis-identifying the problem can lead to misguided analysis. If a health care professional misdiagnoses a problem and assigns treatment, the problem will most likely still exist. Sometimes what is thought to be the problem is only a symptom of the true problem. Consider the following example:

"When our school releases students at 3 p.m., there is a huge traffic problem that at times brings our area to a standstill."

Although on the surface, all of the cars being released at once from the school may seem to be the problem, in research it was discovered that a refinery located on the same major road as the school, recently moved a shift change to 3 p.m. This doubled the amount of traffic in what was already a congested area.

The **Problem Identification Statement** should be a short paragraph that answers the following questions:

- **What is the initial problem?** Define the problem based on personal observations or from reports of others.
- **Why is the study needed?** Provide the importance of action being taken to address the problem.
- **Who is the person, client or organization that has the problem, if applicable?** List the person or group that has the most interest in addressing the situation or solving the problem
- **What form will the solution of the problem take?** Providing map layouts, a written report, and an oral presentation may be sufficient or it may provide a stepping stone to that group to solving a larger issue.

An example of a **Problem Identification Statement**:

The U.S. Department of Transportation wants to identify the most expedient evacuation routes for all U.S. coastal areas prior to the hurricane season this year. More than half of all Americans now live on or near a coast, a major concern for evacuating residents in a timely manner in the event of a natural disaster. Analysis results will be submitted to the USDOT using a written report, and the USDOT will communicate the findings to coastal residents using the USDOT website and evacuation brochures that will be distributed as inserts in coastal newspapers and through local governmental agencies.

Keep in mind that the **Problem Identification Statement** may see many drafts before the most accurate definition of the problem is achieved. Many of the topics discussed in **Project Planning** can undergo changes, particularly after interviewing project stakeholders.

Project Stakeholders

Project Stakeholders include individuals, families, agencies, or organizations that have a keen interest in the project and/or the project findings or outcomes. They can come from a diverse range of backgrounds and have different types of stakes, or interests, in the issue at hand whether they are personal reasons, economic reasons, professional reasons, or even legal reasons. Identifying ALL project stakeholders is essential to the success of a project. All feedback given by a stakeholder – positive and negative - should be considered to complete the project. It is important to take all the time necessary to identify all project stakeholders relevant to the study as well as identifying their stake in the issue.

In the traffic example used in the **Problem Identification** section, a possible list of stakeholders and their interest/stakes could be:

Stakeholder Examples:	Types of Interest/Stakes:
Residents	Personal
Local Employees	Economic & Personal
Students	Personal
Shift Workers	Personal & Professional
Refinery Management	Professional
Area Businesses	Economic
School District Administration	Professional
Parents	Personal
Shift Supervisor	Professional
City Police	Legal
City Government	Economic & Legal
Department of Transportation	Economic & Legal

There are potentially even more stakeholders for this project than listed. Taking the time to look around the area and take an "inventory" of the businesses and assets could help produce a more thorough list. The list, of course, will vary depending on the topic being studied. The importance is to identify and include all stakeholders in the process, no matter what their opinion of the subject matter may be.

Once you identify the stakeholders, it is imperative that you meet with them in some capacity - whether it is an interview with one or a community forum with a group - to gather their input on the matter at hand. Other possible methods could be sending questionnaires to stakeholders, polling them (polls are similar to questionnaires but do not have opened questions), or conducting focus groups. The method or methods selected to do this will be at the discretion of the project manager.

Project Objective
The **Project Objective** involves establishing the project's goals so that all personnel know the focus and scope of the project. To be effective, the project objective needs to be as specific as possible.

Peter Drucker offers this method for defining effective project objectives in "Management by Objectives". Project Objectives should be **SMART**[1]:

- **S**pecific – states what should be achieved
- **M**easurable – establish a way to determine goal obtainment
- **A**chievable – the goal communicated must be obtainable
- **R**elevant – will help provide direction and structure
- **T**ime bound – provides a time frame for obtaining the goal

Consider the following example:

By the beginning of hurricane season 2008, determine the best evacuation routes for coastal areas in the U.S. in case of natural disaster that could be utilized to evacuate the entire population within 48 hours of the disaster.

This objective statement contains all of the SMART characteristics: specific, measurable, achievable, relevant, and time-bound. Writing an objective will require quite a bit of research and must include the stakeholder analysis to properly define the issue. Just as with the **Project Identification**, it may be necessary to revisit the initial **Project Objective** as further **Project Planning** activities continue.

Project Title
The **Project Title** should be a single, concise idea for all of the coordinated activities involved in the project. This title should be descriptive enough to keep all members of the group focused on a common goal. It should be relevant and interesting to those working on the project. Below are some examples to consider:

Drive-Time Traffic Flow Studies in the Detroit Area

Tracking Effect of Mosquito Control Applications in St. Tammany Parish

---

[1] Peter Drucker, "Management by Objectives," 1954.

The **Project Title** sets the mood for the entire project and therefore should not be a hastily made decision. The perfect title will enable those working on the project to focus on a specific goal. This is not necessarily the title that will be used on the written report during the Project Presentation portion of the project.

Geographic Extent

Although the **Geographic Extent,** or **Area of Interest (AOI)**, of a project may seem obvious to you when setting up the project, it is very important to define the extent for all who are involved or who may read the project. It may encompass the world, a single country, a state, a county, a campus, a neighborhood, or a city block.

In the school traffic example explored earlier, the extent of the traffic problem initially involved just the traffic issues just around the school's campus. The extent at that time was less than a few city blocks. After further investigation, it was decided that the route that the refinery workers would drive to get to and from work – including the area where the school is – needed to be studied also. The extent at that point went from a few city blocks to studying the entire city.

Defining the extent keeps everyone involved aware of the specific study area of the project. It also has to be established before the data acquisition phase in the Project Implementation occurs.

Project Feasibility

Can you afford do conduct this study? **Project feasibility** must be determined in terms of not only monetary costs, but also in terms of time, personnel, and material costs involved versus the potential benefits that will be received from conducting the study.

Although a large office complex may want the city to build a pedestrian bridge over a busy road that cuts through the complex, is it feasible to pursue?

After a study like this is conducted it may be concluded that the pedestrian bridge may be necessary or the conclusion may be that installing a traffic light may be more feasible than building an actual bridge.

The feasibility of a project may have numerous constraints that must be considered. Listed below are a few examples:

- **Schedule feasibility** – assigning a certain amount of time to complete the project considering, both, stakeholders' needs and team members' needs
- **Legal feasibility** – researching legal documentation to preserve project validity
- **Political feasibility** – weighing the pros and cons of those involved
- **Technical feasibility** – determining realistic access to the technical resources available for project
- **Organizational feasibility** – assessing if those involved can fulfill the project requirements
- **Financial feasibility** – estimating the costs vs. benefits
- **Functional feasibility** – based on project focus – i.e. is the proposed new building site suitable, is the area too mountainous, does it have limited access, or will the area need to be cleared due to too many trees?

Each of these different feasibility focuses is important to the overall review of the project in terms of its viability. Some may be more relevant to a project than others but each should be evaluated as a precaution during the feasibility assessment portion of **Project Planning**.

Functional Requirements

Compiling a list of **Functional Requirements** involves predicting what you need to conduct your study. More specifically, it involves identifying the following:

- **Data** – buy, acquire yourself, or find for free
- **Equipment** – data collection instruments, vehicles, safety equipment
- **Computer Hardware & Software** – finding computers, appropriate software, printers and other peripheral devices
- **Personnel** - hire new or use existing
- **Supplies** – office space, paper, pens, desks, etc.

In the traffic example used earlier, the type of data you would possibly need would be traffic and pedestrian counts, and road or street files. The street data might be available online for free, but the traffic and pedestrian counts may require buying this information from a third party or hiring someone to collect this data for you. The tools in that same example may be the computers used to perform the analysis, the counters used to count the students and traffic, and the vehicle used to drive to the site and collect the data. The personnel may require that you hire a person to collect the counts or to find the data online and the supplies are general office supplies that you would find in just about any office, but when starting out, will need to purchase.

It is important when compiling this list that it is as extensively researched as possible to identify needed assets. The goal of this step of **Project Planning** is to identify what is needed to successfully complete the project while being mindful of the client's needs. Successful completion of the Functional Requirements step should produce a thorough list of needs to complete the clients request without excessive or unnecessary data. Without an accurate view of the functional requirements, the project may either fail or exceed budgeted time and funds. These items are the ones that you will acquire in the next phase of the project, **Project Implementation**. Stakeholder interviews can also provide a key source of additional information about the **Functional Requirements** that are needed for a project.

Project Design

Once all of these steps have been examined, the **Project Design** can be completed. In a **Project Design**, an outline of the entire project management process can be made and should be written so all members of your team have a clear idea of all of the steps that will take place. Steps from new hiring (if necessary) to acquiring data for a task to creating a map layout to reporting your findings should be listed in the **Project Design**. Although the actual steps at this point have not been implemented, this will give the project manager a way to map out all of the processes and procedures that will soon take place to meet the project goals. Additional steps may be added once project implementation takes place, but the **Project Design** will provide project managers with the ability to see the "big picture" to be able to adequately communicate instructions complete with specific goals to his or her personnel. A complete **Project Plan Summary** provides an easy transition into the **Project Implementation** phase of the **Project Management Model**.

Please keep in mind that any unexpected or new findings in any of the above steps may require that the problem be re-identified. If this is the case, all of the steps involved may have to be revisited and possibly edited. Revisiting these steps will take a fair amount of time to accomplish but will be worth it in the end. For instance, if the location that a business was to build on was found to be residentially zoned, then the process would have to be revisited again.

Scenario:

The city government of Hartsville has not yet seen firsthand how geospatial technology can benefit them. As a geospatial technician, you have been asked to conduct a study of the traffic congestion situation that takes place in the morning hours on Main Street. You have been given six months to conduct your study. Following the **Project Management Model**, your work will begin with exploring the different steps of **Project Planning**, ending with a **Project Plan Summary** as seen on the following pages.

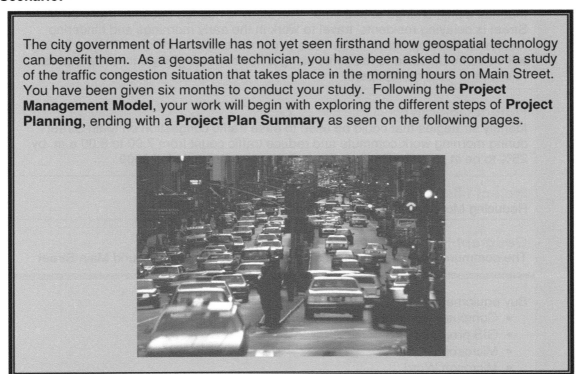

Project Plan Summary

Main Street Morning Traffic Commute Scenario

Project Planning

Problem Identification:
Main Street is heavily congested during morning rush hour traffic due to work commuters, two schools near Main Street convening, and one factory in the area that has a shift change at that time. The study is needed because traffic on Main Street is delaying residents' travel to work in the early mornings and hindering residential developments in town due to current traffic conditions. Project findings will be communicated to residents using a "town meeting" style forum and press releases to local newspapers.

Project Objective:
Identify strategies that could be used to ease traffic congestion on Main Street during morning work commute and reduce traffic count from 7:00 to 8:00 a.m. by 25% to be in place by the time that school commences in fall 2009.

Project Title:
Reducing Morning Commute Traffic on Main Street

Geographic Extent (AOI):
The community of Hartsville, focusing in on a 5 block radius around Main Street

Project Implementation Weeks 1 & 2

Buy equipment
- Computer workstation and Printer
- GIS program
- Microsoft Excel
- Microsoft Word
- Microsoft PowerPoint
- Counting device for measuring school and factory traffic

Order/collect data
- Street network
- Census block groups with population attributes & commute statistics
- Traffic count data
- School population
- Factory population

Project Implementation Weeks 3 -8	<ul><li>Create a base model for the project in ArcMap showing the locations of the schools, factory, and Main Street</li><li>Create maps displaying various traffic flow patterns from:<ul><li>Schools</li><li>Factory</li></ul></li><li>Create maps showing different traffic flow patterns for other times during the day</li><li>Determine if alternate routes are possible for schools or factory workers</li><li>Analyze data from all maps created</li></ul>
Project Implementation Weeks 9 - 15	<ul><li>Develop an outline of the written report to be presented to the stakeholders.</li><li>Document major traffic hot spots as well as peak times for congestion for inclusion in the project report</li><li>Write the written report including the maps created as well as the conclusion derived from the evidence presented</li><li>Create visual aids and any other presentation material for the presentation</li></ul>
Project Implementation Weeks 16 - 20	<ul><li>Conduct final edits on the presentation materials and the written report</li><li>Establish a time and date for the oral presentation of the report</li><li>Community stakeholders will be informed of the presentation date and time through a press release and through a town meeting forum</li></ul>
Project Presentation Weeks 21 - 23	<ul><li>Finalize edits and presentation materials</li><li>Send out invitations to stakeholders</li><li>Create and send out press release</li><li>Presentation of project findings to stakeholders via town meeting forum</li><li>Document feedback from forum participants</li><li>After presentation, present all feedback to City Government of Hartsville.</li></ul>

Conclusion

As is the case in the scenario above, it is very beneficial for the project manager to keep detailed documentation of all of the steps involved in **Project Planning**. This will allow for a smooth transition into the next step of the process, **Project Implementation**. Many times when a situation like this arises, an entire group of people are assigned to help determine what the issue is and the steps can be taken to resolve the issue. Can you see how it would be beneficial to have a group involved instead of having to complete this project as an individual?

All of the steps involved in the **Project Planning** phase - from identifying the problem, identifying stakeholders, establishing an objective, creating a title, defining an AOI, determining feasibility, listing all of the functional requirements, to creating a project plan summary - will take a significant amount of time to examine. Comprehensive **Project Planning** typically results in a better use of resources, a more efficiently executed project, and ultimately a successful conclusion to project activities. The extra time invested in **Project Planning** will save time and possibly resources in **Project Implementation**.

Lesson 1: Project Planning Stakeholder Exercise

Project Planning is an extremely important part of Project Management. In order to ensure that the project meets its intended goal and objectives, thorough project planning must take place. Without adequate project planning, a project may commence only to be halted later due to unforeseen circumstances that proper planning would have probably identified.

One way to help ensure that Project Planning is thorough is by identifying all (or as many as possible) stakeholders who have an interest in the project. By identifying the stakeholders an seeking their input during planning, a GIS project manager can get a more comprehensive view of the problem or situation at hand and be better prepared to manage obstacles along the way.

Directions: Read the following scenarios and list the stakeholders in the area provided along with their interest (personal, economic, legal, and/or professional) and if necessary a brief description of that interest or stake in the situation.

Scenario:
A nationally established clothing company is considering opening a retail location in a town with a population of 50,000. The company targets middle income individuals between the ages of 18 - 45. The company is looking for a location that provides the most convenient access to its target customers.

Stakeholder Analysis List	
Stakeholder	**Brief Description of Interest in Project**

Stakeholder Scenario #2
The smell of natural gas was filling up the largest store in the mall, Stacy's. Everyone in the mall needed to be evacuated. To be precautious, every home and business within a two block radius had to be evacuated.

Stakeholder Analysis List	
Stakeholder	**Brief Description of Interest in Project**

Lesson 1: Project Planning Lesson Review

Key Terms

Use the lesson or glossary provided in the back of the book to define each of the following terms.

1. Project planning
2. Problem identification
3. Project stakeholders
4. Project objective
5. Project title

6. Geographic extent
7. Project feasibility
8. Functional requirements
9. Project Summary
10. AOI

Feasibility study

It's not only the dollar amount it will take to conduct a study when you are considering feasibility. To become more familiar with the different types of feasibility, match the descriptive question to the correct term.

_____ 11. Schedule

_____ 12. Legal

_____ 13. Political

_____ 14. Technical

_____ 15. Organizational

_____ 16. Financial

_____ 17. Functional

a. Are there enough monetary funds to complete the project?
b. Is there enough time to complete the project?
c. Is the terrain of the land a factor?
d. Do we have the right people for the job?
e. Is the land properly zoned for this project?
f. Will we have the community's support?
g. Can we get the equipment we need to conduct this study?

Global Concepts

Use the information in Lesson 2 to answer the following questions.

18. Briefly explain why it is so important to define your AOI.
19. Why is it important to consider all stakeholder's opinions – even the negative ones?
20. Which step in Project Planning is said to be the most difficult? Why?

Let's Talk About It...

Answer the following question and share the responses with your instructor and classmates.

21. The building you are in is going to be expanded to accommodate several more classrooms. Brainstorm with your classmates to create a list of the functional requirements that might be needed to conduct a study of where to expand the building.

Lesson 2:
Project Implementation

The **Project Management Model** (**PMM**) was created to enable those using it to properly plan, execute, and present a project from start to finish in the timeliest, efficient, and most resourceful way. The three major components of the **PMM** are **Project Planning**, **Project Implementation**, and **Project Presentation**.

In **Project Planning**, the problem was identified, an objective was established, a title was given, the geographic extent was determined, the stakeholders identified, the feasibility was studied, the functional requirements were listed, and a project design was created. Now that the groundwork has been set up for the project, it is time for the next step, **Project Implementation**.

Project Implementation involves four important tasks that must be managed;

- **Resource Acquisition**
- **Data Acquisition and/or Collection**
- **Data Processing & Analysis**
- **Map Layout Production**

In order to gain a better understanding of **Project Implementation**, let us take a look at each of the tasks in greater detail.

Resource Acquisition
Resource acquisition involves collecting and acquiring the functional requirements that were identified during the **Project Planning** phase of the **Project Management Model**. The tools necessary to complete the project must be acquired – including computer hardware, geospatial software, and the data collection instruments (such as GPS units, soil or water testing instruments, etc.).

In addition to tools, personnel must be hired or current workers assigned with regards to the tasks and responsibilities they will be responsible for throughout the project. The type of personnel that are to be hired will depend on the subject matter being studied. Finding someone internally may be the least expensive way to acquire assistance for a study since additional pay would not be an issue. However in some studies, specialists may be needed. For instance, if conducting a study of gas lines in a community, a local oil and gas person may need to be hired. Some other examples may include:

- **Researchers** – paralegals, archeologists, historians
- **Analyst** – GIS or topic specific
- **Data collectors** – Surveyors, GIS professionals, students
- **Trainers** – GIS trainers or procedural trainers
- **Engineers** –Structural or Electrical Engineers

Whether hiring new team members or training current staff; the process of doing these procedures can take up a considerable amount of time. Take a second to think about the procedures that you have to go through to hire someone - put ad in the paper or post it on the Internet, wait for response, bring in interviewees, hire the one best fit for the project. It might take as little as two days or as much as two months to find the right person. This one decision alone could possibly push back the project completion date and could cause the project to come in over budget.

Other resources that might be acquired during **Project Implementation** may include the basics – such as office supplies, desks, computer supplies, and binders. There may be other supplies that are a little more challenging to find around the office or at your local office supply store. As a project manager, if you have to acquire a soil PH kit, for instance, would you know where to purchase one? Not sure? This is one of the many duties that a project manager may have to take the time to research then budget for and set up a way to acquire it. Project managers must develop the foresight to budget in extra time for unforeseeable happenings such as these covered in this section.

Data Acquisition

With the help of the Internet, many different types of data are now available with a few strokes on the keyboard and a few clicks of the mouse. Various shapefiles ranging from historical markers in your county to precipitation information for your region, to imagery showing your local neighborhood or even your entire state can now be accessed on the Internet. When acquiring geographic data, it is crucial to use valid, credible sources. The federal government provides an extensive source for geographic feature data and imagery with many of their sources providing information for free. A sampling of some of these sites is provided below.

Federal Agency	Link	Type
United States Geological Survey	http://www.usgs.gov/	*Science org focusing on geospatial info, biology, geography, geology, & water*
National Oceanic and Atmospheric Administration	http://www.noaa.gov/	*U.S. Dept of Commerce site w/info about weather, satellites, oceans, and more*
National Geodetic Survey	http://www.ngs.noaa.gov/	*Provides info for transportation, communication, mapping, charting, and more*
TerraServer	http://www.terraserver.com/home.asp	*Sells aerial photos, satellite images, and USGS topo maps*
U.S. National Park Service	http://science.nature.nps.gov/nrdata/	*Provides data & metadata files of National Parks*
U.S. Environmental Protection Agency	http://www.epa.gov/	*Provides links to downloadable data pertaining to environmental issues*
U.S. Census Bureau	http://www.census.gov/	*Provides many types of*

		downloadable census data files
GeoData.gov	http://gos2.geodata.gov/wps/portal/gos	Portal providing links to various geospatial data sites

Keep in mind that there are many other sites available from the federal government. Data needs will determine the best site to be used for a project.

On a state level, many states now have organizations established with the purpose of providing free or low-cost geographic data that is readily accessible by the public. In addition to these organizations, there are various public and private companies that also provide geospatial data and imagery. The following is a list of just a few of these organizations.

Acronym	Organization	Link	Type
	GeoCommunity - GIS Data Depot	http://data.geocomm.com/	Portal that provides links to various types of geospatial data both free and for sale
	Digital Globe	http://www.digitalglobe.com/	Provides digital images and related services
ESRI	**Environmental Systems Research Institute**	http://www.esri.com/	Provides numerous types of geospatial data and services
PASDA	**Pennsylvania Spatial Data Access**	http://www.pasda.psu.edu/	Provides aerial photos, topo maps, and state-wide data for the state of Pennsylvania
TNRIS	**Texas Natural Resources Information System**	http://www.tnris.state.tx.us/	Provides aerial photos, topo maps, and state-wide data for the state of Texas
NJGIN	**New Jersey Geographic Information Network**	https://njgin.state.nj.us/	Provides aerial photos, topo maps, and state-wide data for the state of New Jersey

There are, as is the case with the federal level sites, many other sites that can be found on the Internet. These listed have been proven to provide reliable data. Caution should be taken, however, when searching on the Internet to find data relevant to a study. Because the data is available on the Internet does not guarantee it is from a reliable source. Just as a computer is only as good as the information entered into it, a geospatial project will only be as good as the data that is used in it. Some data may be subject to significant error and may not be as suitable as other sources of data. Error comes from a variety of sources including human error and instrumentation precision (for data collection tools). Some of these sources will provide geographic data with data error or data accuracy measurements while others will not.

Counties of New Jersey, New Jersey State Plane NAD83
Map Preview | View Metadata
Download: Personal GDB | Shapefile

Example of downloadable data from NJGIN

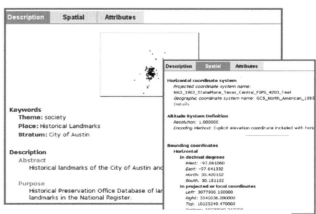

Whenever data is created and stored in ArcMap, it has a blank **metadata** file attached to it that the creator can complete. What is metadata? Simply put, **metadata** is information

---

about data. A **metadata** file allows the creator to put their own "virtual signature" on their own work to help validate the file. Typical information in a metadata file will include who created it, when was it created, where was it created, why it was created, how it was created, and contact information for the creator, and much more. Metadata files may also contain information about the projection or coordinate system used when the data was created. When data is downloaded to be used in a project, it is crucial that the user study the metadata to verify not only the validity of the source but also to determine if the date the data was created is recent enough to be applicable to the study.

One other concern when gathering data from various sources on the Internet is that data does not always come in a neat, easily usable format.   The Federal Geographic Data Committee (FGDC) was established to promote one nationally recognized, standardized publishing format for all geospatial data created. Unfortunately not all data follows FGDC format yet. For now, data might come set up in a standard shapefile or it may come packaged in a table or spreadsheet type format. The metadata with data files may meet FGDC standards or it might not exist with the data at all. It is important that the user be aware of what they are getting and what the source of their data is.

Regardless of the source or sources of the data acquired, it is imperative that the project manager properly manages his resources, such as time spent acquiring the data or imagery as well as the monetary value of the product purchased, in mind.

Data Collection
If the data needed for a study is not available or the source for found data is unsuitable for the project, it may be necessary for the project manager to collect their own data. For instance, if an insurance company needed to know how far each of the houses they insure are from fire hydrants, a person could simply go to the neighborhood and collect data on the hydrants using a GPS unit. That information could be easily transferred to mapping software for plotting the locations of the hydrants on a map.

Depending on the project, data collection may range from as simple as collecting coordinate data yourself with a GPS unit, to hiring a specialist to collect and test soil or water samples. Both project managers as well as team members alike should create a metadata file when building a data file using FGDC standards. This will allow them to fully document what they did, why they did it, and much more to give as much creditability to the data collected as possible.

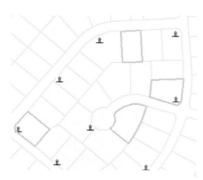

With data collection, as with other steps involved in Project Implementation, it is important to consider the costs associated with data collection; not only in terms of money but also in the time it will take to gather and process this new data.

Data Processing & Analysis
Data processing and analysis finally involves taking the data collected and plugging it into a geospatial software program where it can be analyzed. To continue with the insurance scenario from earlier, a map can be made showing where each of the land parcels that contain homes that the insurance company insures is located. Next the GPS data collected with the coordinates of the hydrants could be added and then displayed on that same map so that the insurance company could see where the hydrants are in relation to the houses that

they insure. Why is this important? The homes that are closest to fire hydrants could possibly be eligible for lower rates.

Although there are other programs that can be used for geospatial analysis, for this course you will use ArcGIS, a software program from Environmental Service Research Institute (ESRI) to conduct your study. ESRI is a leader in the geospatial software industry and has been providing software and educational products to the world for over 30 years.

Map Layout Production

Once all of the analysis is complete, map layouts can be created to communicate the project findings. In some cases, such as the insurance scenario discussed above, only one map layout may be necessary. In others, it may take a number of layouts to complete the study. An example of this could be studying a homeland security issue where there may be a need to create several maps to show different scenarios in a geographic area.

Conclusion

Project Implementation is the second phase in the **Project Management Model** after **Project Planning**. It involves **resource acquisition**, **data acquisition and/or collection**, **data processing and analysis**, and **map layout production**. **Resource acquisition** involves acquiring the necessities to complete the study as outlined in the functional requirements section. When acquiring data from the Internet, it is important to find sources that provide valid, credible data. It may be necessary to create or collect your own data for a study. It is important when obtaining your own data to keep in mind the potential time and monetary costs that may be involved. Once all the data needed for a study has been gathered, it will need to be put into a mapping program such as ArcMap to be analyzed. To communicate the findings of the study, a map layout can be made complete with all necessary map elements. There may be a need for more than one map layout to be made depending on the study being conducted.

Lesson 2: Project Implementation Data Needs Exercise

In order to ensure the validity of the GIS project, it is imperative to use only existing data that is obtained from a credible source. One of the ways to ensure that data is valid is by using data that follows standards and guidelines set forth by the Federal Geographic Data Committee (FGDC). This organization established the National Spatial Data Infrastructure (NSDI) that is an effort to work with the public and private sectors to encourage the efficient and effective sharing of geographic data consistent in format, accuracy and dependability. The FGDC hosts a clearinghouse that allows users to search for geographic data from a variety of reliable participating data repositories (*www.fgdc.gov*)

Data does not always come in a neat, easily usable format. You may have to manage the data in a spreadsheet program and use various techniques to make it usable in a GIS software program. However, this exercise is designed to have you to think about the different types of data that could be used in a GIS project and possible sources for that data.

The following organizations are examples of good sources of data that can be used in local, state and national GIS projects. A simple Internet search of these agencies and organizations can direct a GIS user to websites where data can be searched and downloaded, oftentimes at no cost.

- *County and Municipal Clerk Offices*
- *County and Municipal Offices of Emergency Services*
- *County Tax Collector*
- *Statewide GIS Web Libraries*
- *US Census Bureau*
- *Geographic Names Information System (GNIS)*
- *FGDC Clearinghouse*

Review the scenarios from Lesson 1 and identify data that may be used in research to make decisions about the scenario topics. Include any data that could effectively support the research involved in the scenario. List these data needs below, as well as possible sources where this data can be obtained.

Scenario #1 Title:	
Data Needs	**Possible Source(s)**

Scenario #2 Title:	
Data Needs	**Possible Source(s)**

Lesson 2: Project Implementation Lesson Review

Key Terms
Use the lesson or glossary provided in the back of the book to define each of the following terms.

1. Project implementation
2. Resource acquisition
3. Data acquisition
4. Data collection
5. Data processing & analysis
6. Metadata

Global Concepts
Use the information in Lesson 3 to answer the following questions.

7. Explain how Project Implementation differs from Project Planning.

8. List at least three different categories of resources that will need to be acquired during the project implementation phase.

9. Why is it important to find credible sources when acquiring data?

10. Is all data found on the Internet free? Explain your answer.

11. Once all of the data is collected, what is the next step in the Implementation process?

12. Who creates a metadata file?

Surfing for Info...
One important role of a Project Manager will be to find shapefiles and imagery to be used in the study. A Project Manager should also research the cost involved of acquiring imagery. Use the Internet to find sources for the following requests.

13. Go to http://www.terraserver.com/home.asp. Under Image Search, type the address of your current location. What is the price of a 3000 x 3000 px image?

14. Go to https://njgin.state.nj.us/ to view the New Jersey Geographic Information Network site. Click Data Downloads on the left side of the screen to display the different data sets available for download. This site is set up so that you can click on **Map Preview** (to see how the file will appear on the map), **View Metadata** (to get detailed information on the purpose, who created it, etc.), and you can also **Download** the shapefile for that data. Take a moment to look at some of the metadata files.

Which metadata file contains this image?

Acronym challenge

This lesson discusses numerous agencies and organizations that you will become familiar with as you work through this series. Match the letter beside each description to the Acronym which best fits it. You may have to use the Internet to find the correct answer.

_____14.	FDGC	A.	Contains numerous data files and more for a state and is maintained by its state's office of information technology.
_____15.	ESRI	B.	Contains numerous data files and more for a state and was originally established as a water-oriented data bank.
_____16.	NSDI	C.	Site for a company that provides GIS data and services.
_____17.	TNRIS	D.	Established to promote one nationally recognized, standardized publishing format for all geospatial data created.
_____18.	NJGIN	E.	Online portal to serve as a "one-stop" for geospatial information

Let's Talk About It...

Answer the following question and share the responses with your instructor and classmates.

20.   Explain why it is important to study a file's metadata.

21.   Is it always necessary to have only one layout per study? Explain your answer.

Lesson 3:
Project Presentation

The **Project Management Model** was designed to help project managers successfully plan, prepare, and execute all necessary tasks to provide their stakeholders with a complete and thorough study of a particular problem or issue. From the **Project Planning** phase where a focus is set and plans are put into place on how to complete the project, to **Project Implementation** where the necessary materials are obtained and analyzation takes place, to this final phase, **Project Presentation** - each part is equally important to the success of the project.

In **Project Presentation**, the findings for the entire project are packaged in order to communicate them to the stakeholders. The project manager must be able to effectively communicate the ideas to the intended audience or this one phase can cause the entire project to fail. Two popular presentation methods that are used in the **Project Presentation** phase of the **Project Management Model** are the **written report** and the **oral presentation**. Oftentimes though, combinations of both methods are used to communicate project findings to stakeholders. The decision for the method of delivery is usually made well before the project is underway. The project manager must keep in mind the intended audience that will be receiving the final report when constructing the report; whether it is in written or oral format. The wording in a written report intended for a group of GIS professionals will not be written the same as one intended for a varied group of stakeholders living in a certain neighborhood in a city.

The Written Report
The **written report** serves to provide detailed information about all aspects of the project. The client or manager that has requested this geospatial study and you, as a project manager, will need documentation of the activities you perform for future references. Project Managers and their teams and the companies/agencies/third parties that they represent undertake a geospatial project and all share a unique stake - accountability. This undertaking provides a service or information that will benefit the community at large, and at some point, now or in the future, a stakeholder may have a specific question that needs to be answered. A detailed written report documenting your methods and findings allows questions to be answered sufficiently and supports validity of the project. The written report provides the opportunity to communicate:

- Project findings & analysis,
- Any project deviations, and
- Recommendations for further study.

Writing a report for a geospatial study is not very different from writing a paper for other research. The project manager will decide if the report should be written as a persuasive report

or as an informative one. Project managers must also decide what information is included in a report and which writing format to use to write that information. A well constructed report allows stakeholders to be able to look at the table of contents and have the option to read the entire report, or focus in on the pages that directly pertain to them.

In the Project Management Model, when composing the written report, there are several major questions that the project manager needs to take into account to make certain that the report is complete. Clients need to be told the results, the methods, and the reasoning for these methods. Over the duration of the project managed using the Project Management Model, clients will not necessarily be aware of all the work done during planning and implementation; however, these activities need to be documented. Answering the following questions can help insure that clients are informed of all of the project activities.

> 🌍 **Why did we use the methods we used?**
> 🌍 **How did we find the answers?**
> 🌍 **What did we find out?**
> 🌍 **What do we do now?**

Let's take a more detailed look at these questions - discussed in the order that they occurred.

🌍 **Why did we use the methods we used?**
> The written report is the chance to share findings with stakeholders, clients, and other interested parties. But in addition to findings, the written report should also show the process that was used to arrive at those findings. Some clients and stakeholders may only be interested in results, but stakeholders who have a financial investment in a project may want documentation to justify costs. Explanation of planning methods can illustrate effective project management and maximize the value added.
>
> The report should cover the major project defining activities of the planning process that took place. These activities are mostly found in the first three steps of Project Planning. Problem identification is important to stakeholders because this step is the basis for the entire project. Discussion of stakeholder analysis is important because it assures clients that all interested parties' issues have been considered in the project. The project objective(s) show stakeholders goals that have been identified using their input assuring that their needs are the focus of the project.
>
> The rest of project planning defines tasks and implementation. Some of these may or may not need to be discussed. Project managers must decide between what information is essential for stakeholders reading this document and information that is trivial and distracts from the written report's goal. The client will be interested in the geographic extent of the study but may not be concerned with the detail of office supplies and hiring guidelines, for example. If there were deviations in the plans contrary to the expectations of stakeholders such as needing additional high resolution imagery for analysis or expanding the geographic extent to include a county instead of just a city, then the stakeholder needs to be aware of the situation.

How did we find the answers?

The report should include a narrative of the process of acquiring and utilizing the functional requirements to complete the implementation phase and meet project objective(s). Descriptions of implementation activities serve to validate results to stakeholders requesting the project in the event that the results presented are subject to questions about its reliability. By detailing data processing and analysis, the findings are supported which gives stakeholders confidence in your research. This would also be the place to document the reasons behind the final map layout (or the various layouts) created.

In larger companies, it may have been necessary for more than one team to be assigned to a study. Documenting how the work was delegated shows efficient time management - while one team is collecting data, another may be creating a base map, and another may be acquiring high resolution images. In the end, however, they will have combined their efforts for the final analysis for the study.

Other information such as the detailed process of interviewing personnel for the study may not be considered relevant or appropriate for inclusion in the written report.

What did we find out?

The report should summarize the results of the project analysis. This is what the stakeholders want to know the most, and depending on your client or stakeholders, this may need to be included near the beginning of the report. In summarizing results, it is necessary to compare actual outcomes to anticipated results. Any deviations in the initial plan should also be discussed in this section – for example, finding out after interviewing the stakeholders that the traffic problem was not just confined to a certain residential area, but to the nearby school as well. In cases with deviations, often times certain factors cannot be mitigated or anticipated. That is expected, but these factors or changes should still be explained as deviations to document how the situation is overcome and used to positively impact the project as a whole.

In the **written report**, the map layouts created in the **Project Implementation** stage are included to provide a graphical representation of the information documented in the text. With GIS projects, your results will be illustrated with maps, so a copy of all layouts should remain with the written report in some usable form.

What do we do now?

Also of interest to the stakeholder will be ideas on what they should do next. This part of the written report allows for the project manager and team to express any recommendations for further study or project enhancements. For example, documenting where crosswalks could be placed on a busy thoroughfare might be the answer to a traffic/pedestrian issue, but recommending where additional sidewalks could be added for additional pedestrian safety could be something the stakeholder could consider for future endeavors.

Each of these components is crucial to creating a well designed and documented report. The order that these questions are answered in the written report is up to the discretion of the project

---

manager. It is important that the project manager gear the written project to the appropriate level of its intended audience. This is your chance to fully document all work, even the smallest of details, appendices can be added to the end of the report. Appendices are excellent for adding data and descriptions that may needed for later reference.

The overall benefit of using a written report to communicate ideas to stakeholders is that a written report will contain many details about the project. Since the details are documented, they can be referenced for future projects or to simply validate your findings years after completion. A detailed report like this may only go to a select group of stakeholders such as the requesting client, upper management of the company, or any other stakeholder who might require detailed documentation of your study. The project may also lend itself to be presented in a mass media setting.

Communicating Findings in Print Media
In a community project that includes many stakeholders from various industries or backgrounds, especially when the general public is considered a stakeholder, additional means of written communications may be necessary to distribute the findings to the stakeholders. These stakeholders, since they are not accountable for the analysis, may only need to know the findings. In this case, some examples of "community forum" avenues for providing written communications include:

- Newspapers
- Community newsletters
- Relevant magazines
- Association journals
- Websites

Depending on the mode of communication used, there may be a limited amount of information that can be given at one time. Rarely, if ever, would the full written report be submitted to print media. Whereas print is a valid medium for reaching many stakeholders in a community at one time, they are limited in the amount of space available to present the findings, especially one with numerous pages and layouts. As a project manager, you and your team members should create articles presenting your findings that cater to both the audience that will view it and the medium that will be used. For example, if you and your group were commissioned to do a highway study, you would create a variety of articles that would cater to the specific audiences of the journals or trade magazines that you chose for publication. With other modes of written communication, it may be possible to provide at least a summary of the written report with information about where copies of the full report are available for review. With a school district, for example, a summary of a written report may be included on the district's website with information that copies of the full report may be available to stakeholders at the school district's central office.

Oral Presentation

Another possible way to communicate the results of a study is through an oral presentation. An oral presentation allows the project manager and team the opportunity to present their findings directly to their stakeholders. Just as with the written report, it is important for the project manager to take into consideration the needs of the audience that the material will be presented to when preparing for the oral presentation.

In an oral presentation, time is usually a factor, and as a result of that time factor, only the main ideas are covered. This of course contrasts the written report which documents not only the main ideas but every detail about how the data was gathered, analyzed, etc. In an oral presentation, the stakeholders will usually want to focus in on what the results or findings of the study were.

- **Project Findings & Results Section –** provide the stakeholders with the results of the data analysis using visual aids such as charts, graphs, or maps created during Project Implementation. The sizes of these visual aids will depend on the number present and the forum used.

However, in addition to these points, the project manager may also want to include a few additional points to ensure that the findings in the project were effectively communicated. These are:

- **A Summary of the Project Plan** – providing the stakeholders with the project objective, the geographic extent (or AOI), and plan of what they set out to accomplish

- **Information on Project Implementation** – provide the stakeholders with how the data was acquired or collected, how it was analyzed, and what types of output were created from the analysis

- **Deviations from Initial Plan** – provide stakeholders with any deviations from the original plan (i.e. an additional stakeholder was added to more thoroughly explore the issue)

- **Conclusions & Recommendations for Further Study/Analysis** – provide the stakeholders with conclusions and recommendations that the project manager and team derived from the information studied

Behind every good oral presentation is structured written documentation that backs up all points and any visual aids used in the presentation. This may or may not be something that the audience will ever see, but needs to be written so that there is ample documentation of the findings as well as any sources used to get the findings. For example, if a stakeholder heard your presentation and wanted to use a data source that you had used or a line you had quoted, that information would need to be available. When conducting an oral presentation, presenters should be careful not to read entirely from any written documentation to their audience but to summarize the reports, hitting all of the major points of their findings. They must also always keep the composition of their audience in mind when presenting – a group of farmers will not need as much background information about land and soil types as a group of banking professionals deciding where to place their next branch.

After the oral presentation, time should also be allotted for a question and answer session so that all stakeholders have ample time to fully understand the conclusions from the project and be able to discuss any issues on the information presented.

Communicating the Oral Results
An advantage of conducting an oral presentation over a written presentation it that the oral presentation gives the project manager and team an opportunity to present the project and project findings directly to the stakeholders thus allowing for quick feedback. Depending on the stakeholders and the study, it could be presented to a few in a small office, to a medium sized group in a classroom, or for larger projects, in a community forum such as a town meeting, council meeting, or convention center. Regardless of the forum, after presenting the findings directly to the stakeholders, the project manager and team  are able to get immediate feedback from them. This feedback could be used to conduct future studies relating to the project.

Conclusion
Deciding on the best way to deliver the message to the stakeholders will require some diligence and should be considered by the Project Manager in the **Project Planning** phase. The following are some very important points to remember:

- Written reports allow for much more in depth discussion on all parts of a report.

- Oral presentations are confined by time parameters but allow for immediate stakeholder feedback.

- Regardless of the type of report that is produced, the Project Manager should keep his intended mind when preparing a written report or oral presentation.

- Successful planning and implementation of a project could fall short if adequate project presentation is not conducted.

Lesson 3: Project Presentation Community Forum Exercise

The community forum in whatever form it takes – from newspapers and radio announcements, to public rallies and town meetings – can be a powerful instrument to use to both gather information for the project from stakeholders and to communicate project findings to stakeholders and the community at large.

In the scenarios you read in Lesson 1, you may have become aware of some media that project managers and organizers might use for these purposes. You may have also thought of other sources of community forums that could be used as communication avenues that would have enhanced the project further. For this activity, you should compile a list of media outlets/community forums used for the project and suggest tactics that could be used to provide project findings to stakeholders and the community.

Scenario #1 Title:
Proposed Media To Use To Communicate to Stakeholders and/or Communicate Findings:

Scenario #2 Title:
Proposed Media To Use To Communicate to Stakeholders and/or Communicate Findings:

Lesson 3: Project Presentation Lesson Review

Key Terms
Use the lesson or glossary provided in the back of the book to define each of the following terms.

1. Project presentation
2. Written report
3. Oral presentation

Global Concepts
Use the information in Lesson 4 to answer the following questions.

4. List the four questions that should be covered in a written report.

5. Name some of the forums that can be used to communicate the results of a written report.

6. For most stakeholders, what are they most interested in hearing or reading about in a report?

7. What is one major advantage of presenting your findings in an oral report rather than a written report?

8. List some of the different sized forums used to present an oral report and discuss how each is beneficial for that venue.

Let's Talk About It...
Answer the following question and share the responses with your instructor and classmates.

9. For a report to be well documented, there are four questions that need to be answered in detail. In what order do they need to be answered in the body of the written report? Why?

10. Should a project manager spend a lot of time focusing on his intended audience when writing a report or preparing an oral report? Why?

11. Why is it important to discuss deviations from the initial project plan?

12. Compare written reports versus oral reports. Name one pro and one con for using each one.

Introduction to GIS Concepts

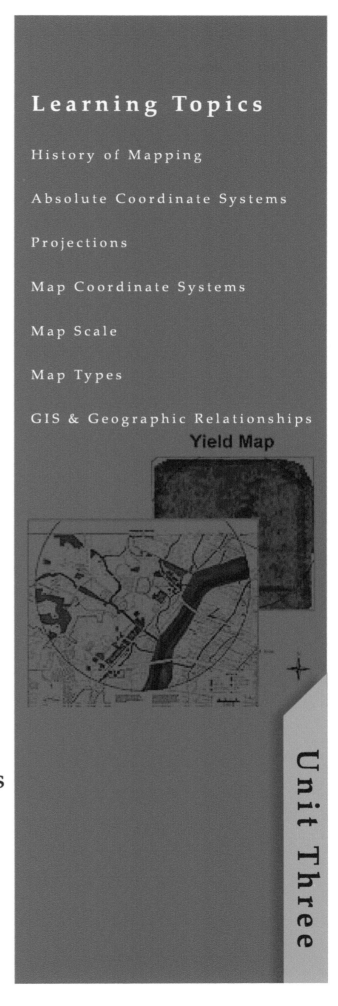

Yield Map

Introduction to
Geographic Information Systems
and Remote Sensing
Concepts

Unit Three

Lesson 1:
History of Mapping

Have you ever had to ask someone for directions to get to a certain location? If you asked three different people, you could possibly get three different answers to lead you to your destination. If you have a map, you can visualize the route yourself. Maps have been essential to navigation since exploration and curiosity caused humans to ask "Where is that and how do I get there from here?" There is evidence that proves that maps have been around for not just hundreds, but thousands of years. They have been created to show where the best hunting places were, where the most fertile ground was to plant in, the best path to take to avoid deep water crossings, the best route to hike or bike in a park, and so much more. Have you ever stopped to think about maps? Where they originated and why they became important? Maps provide the foundation for GIS. In order to fully understand GIS, you have to have a good understanding of maps.

What are Maps?

A **map** is a visual representation of an area shown on a two dimensional surface. **Maps** typically show places from an **orthogonal viewpoint**. **Orthogonal viewpoint** simply means viewing something from above. Many call this the "bird's eye view". Oftentimes **maps** are used to find where a place is or to determine how to get from one place to another. Although these are two very popular usages for **maps**, there are many other types of **maps** that are created everyday for a variety of uses. **Maps** can be divided up into three basic categories:

- **Political maps** - shows how humans have divided the Earth
- **Physical maps** - shows land and water formations on the Earth's surface
- **Cultural maps** - shows patterns of demographic data

These three broad categories cover many different types of **maps** with various purposes. **Political maps** can show information like state, county, and city boundaries (depending on the scale of the map). **Physical maps** will show information like where the peak of a mountain range is to the lowest point of the valley below that peak and other features created by nature. **Cultural maps** provide a means for showing demographic information like where certain age groups or where people with certain income levels live.

This image focuses in on the United States. Which type of map do you think it is?

Within those categories **maps** can be produced that are general **maps** or ones that contain more of a specific cause or theme. Some of these will show everyday, nonspecific general items such as cities and roads. Others will have a more specific use such as a map showing the locations of a fast food chain in a city or a detailed map of a local zoo. These types of detailed **maps** are called **thematic maps** and will be covered in more detail later in this unit.

Color coded, thematic map of a zoo displaying locations of facilities, exhibits, exits, and service locations.

A single location, like a large city, may contain many features that describe it and exist within it. Making one single **map** to represent all of its features would be too difficult to read. Picture a **map** of New York City showing the locations of theaters, historical markers, restaurants, streets, stop signs, fire hydrants, laundry mats, bridges, tourist attractions, television stations, toy stores, and hotels – all on one **map**. It would be confusing and cluttered, wouldn't it? Fortunately, we have the ability to create **maps** that show specific topics such street maps, tourist attraction maps, maps showing utility lines and so on. Each of these different types of **maps** shows specific kinds of information.

The Beginning of Mapping

There is evidence that **maps** were some of the earliest communication tools used by ancient man. Primitive drawings in caves showed the direction to fertile hunting grounds and simple **map** etchings have been found in sandstones. Ancient Babylonian **maps**, such as the example on this page, are drawn in clay tablets that depict the Earth as a flat circular disk. Another culture that created **maps** was the Chinese. The ancient Chinese created advanced **maps** that were very detailed in comparison to other known ancient **maps**.

Babylonian Clay Tablet

Ancient Chinese Map

How Are Maps Used?

When you think of why people use a **map**, typically think they are used to find the location of a place. Where is Rancho Cucamonga, California? Where is Mount Rushmore? You can also use a map to determine how to get from one place to another, how many miles it may be, or how it will take you to drive there. How do you get from St. Louis, Missouri to Austin, Texas? Navigation is not the only other use for **maps**. You can also use a **map** to show geographic features are distributed over an area. Why are many large cities built near water? How far inland from the gulf coast have fire ants migrated? **Maps** have

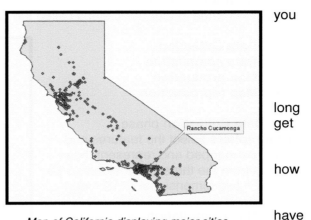

Map of California displaying major cities, highlighting Rancho Cucamonga

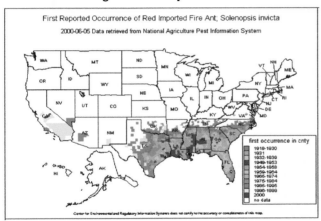

Map showing where and when fire ants were first discovered in the United States.

also been created to help visualize geographic relationships that exist among geographic features. Is there a direct relationship between areas in Washington, D.C. with the highest education levels reside versus the areas where those with the highest income levels reside? The location of manmade and physical entities on the earth are most often not incidental. Why are many of the older cemeteries found near streams or rivers?

Some reasons maps are used:

- **To locate places on the earth**
- **To show patterns of feature distribution**
- **To discover geographic relationships that exist**

Although the reasons that **maps** are used will vary from situation to situation, every map is made for a purpose.

What is Cartography?

Cartography means "the science or art of making maps"[1] and **cartographers** are people who make those maps. **Cartographers** followed largely the same steps GIS Analysts use to create maps: keeping in mind their audience when creating a **map** and using proper layout, symbols, and colors that will appeal to that audience. It is important to realize that without **cartography**, there would be no **maps**. Without **maps**, there would be no GIS!

_____

Cartography. (2009). In *Merriam-Webster Online Dictionary*. Retrieved February 26, 2009, from http://www.merriam-webster.com/dictionary/cartography

The Cartographic Process

Maps are created in four basic steps called the **cartographic process**:

- **Data collection**
- **Data compilation**
- **Map production**
- **Map reproduction**

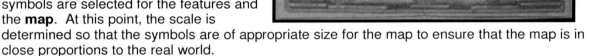

In the **Data Collection** phase, **cartographers** select the features and area to be mapped and represent those features from the three-dimensional (3-D) world in a two-dimensional (2-D) map.

During **Map Compilation**, appropriate symbols are selected for the features and the **map**. At this point, the scale is determined so that the symbols are of appropriate size for the map to ensure that the map is in close proportions to the real world.

During **Map Production**, the first map is created. Keep in mind that this step is the same regardless of the medium used. **Map Production** is much more efficient and much easier today than it was when clay tablets were used for maps, but those same historical maps were created with similar thoughts that you will have a GIS user.

Map Reproduction involves making copies of the original map. With early maps, a single representation may be all that was created. Gutenberg's invention of the movable type made reproduction of maps and other texts possible. Imagine having to carve or draw numerous copies of a map by hand.

Examples of movable type plates – before this invention, books had to be reproduced by hand

As you may have noted, the **Cartographic Process** has similar steps to those found in the **Project Management Model** you studied in Unit 2. There are a set of procedures in creating a **map** from collecting, compiling and producing the **map** similar to some steps in **Project Implementation.** In **Map Reproduction** there are steps that are similar to those found in **Project Presentation**; however, the scope of all these activities is to create one single map as opposed to an entire project which calls upon additional skills and steps to answer a question. The **Project Management Model** allows you to incorporate the **cartographic process** into a variety of coordinated activities.

Map Production

Map production and **reproduction** has changed drastically over the years. The earliest maps, of course, were made by hand. **Reproductions** of these early **maps** were also meticulously made by hand. Eventually, movable type, copper engravings, wooden engravings and other methods were used to reproduce **maps**. The time that it took to produce one good for **reproduction**, however, still took a fair amount of time.

Copper engravings used to reproduce maps.

Today, **map production** and **reproduction** are quite different. Computers and mapping software have provided us with the means to create data, save data, import data from various sources, and make something like a general area map or a specific, thematic map of a certain area, all with a few clicks of a mouse and a few strokes on the keyboard. **Reproduction** also has been streamlined with the aid of color printers that can print maps from stamp size to wall size in less than couple of minutes.

Plotter printers are good for making large sized maps.

Cartography Timeline Exercise

Create a timetable illustrating significant events or periods in cartographic history. Your timetable should include at least twenty (20) events or periods with a brief description of the events or periods included and their significance to the development of cartography. These descriptions do not have to be included on the timeline itself, but can be included as footnotes or some type of supplementary documentation. Use library resources, geography textbooks, or the Internet as sources for this information. Be sure to include the following people or entities in your timeline:

- Claudius Ptolemy
- Peutinger Table
- Thales of Miletus/Anaximander of Miletus
- Aristarchus of Samos
- Mappa Mundi
- Fra Mauro
- Johannes Gutenberg
- Martin Behaim
- Christopher Columbus
- 1507 World Map by Martin Waldseemüller
- Abraham Ortelius
- Invention of the computer

You may either draw the timetable by hand or use a computer for any or all of the assignment. When you have finished, turn the timetable in to your instructor.

Map Categorization Exercise

There are three broad categories that cover many different types of maps; political, physical, and cultural. In this exercise, identify which of the three types of maps with the description or depiction supplied.

		Map Description		Category
_____	1.	A map of the counties in Tennessee	A.	Political
_____	2.	A map showing the voting districts in Harris County, Texas	B.	Physical
_____	3.	A topographic map of Stone Mountain, Georgia	C.	Cultural
_____	4.	A map showing depth contours of the Gulf of Mexico		
_____	5.	A map showing where the most populated cities of the world are		
_____	6.	A map showing where the least amount of individuals aged 65 and over live		
_____	7.	A map of the La Brea Tar Pits		
_____	8.	A map of all the rivers in the United States		
_____	9.	A map of the United States showing the average summer temperature		
_____	10.	A map showing the school districts in San Bernardino County, California		
_____	11.	A map showing the local lake		

Continue the process using the maps given...

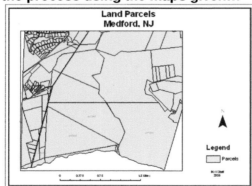

_____12.

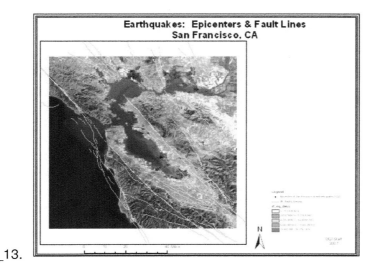

_____ 13.

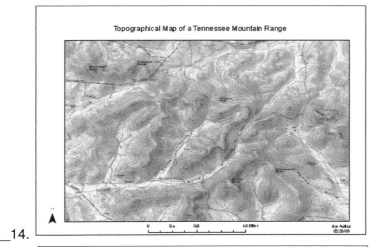

_____ 14.

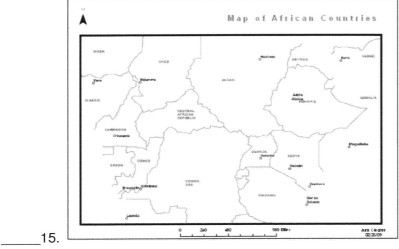

_____ 15.

Lesson 1: History of Mapping Lesson Review

Key Terms
Use the lesson or glossary provided in the back of the book to define each of the following terms.

    1. Map
    2. Orthogonal viewpoint
    3. Political maps
    4. Physical maps
    5. Cultural maps
    6. Thematic maps
    7. Cartography
    8. Cartographers
    9. Cartographic Process

Global Concepts
Use the information from the lesson to answer the following questions. You may need to answer these on the back of this page or on your own paper.

10.   Maps can be categorized into three basic types. List and describe them.

11.   Identify three reasons that maps are used and give an example for each.

12.   The cartographic process contains four basic steps. List and summarize each in the order they are performed.

Let's Talk About It...
Answer the following question and share the responses with your instructor and classmates.

13.   How is the cartographic process similar to the steps involved in the Project Management Model? How is it different?

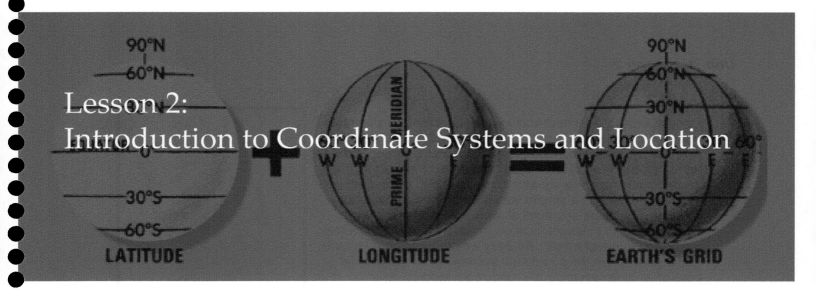

If you were to describe to the person next to you where you were, how would you do it? Would you tell them that you are four feet from the south wall, twelve feet from the door, etc.? If you needed to tell someone in another country where you were right now, would your description be any different (regardless of the language used)? This description may be a little more involved in that you may describe it by saying you are in Ms. Schneider's room, in Beeville, in Bee County, in the state of Texas, and in the United States of America. To give an even more descriptive location, you may include longitude and latitude coordinates. The ability to let someone know of your location can be extremely important. Maps are an invaluable tool to many who travel, but what if you were sent to help out in a coastal town after a hurricane has hit it? If all of the street signs and significant landmarks were no longer there, how would you know where to navigate to find a destination?

Spatial Location & Reference

One way to describe spatial objects or features is by their **absolute location**. An object's absolute location is acquired by using a referenced coordinate system to define a fixed point in space. A coordinate system is a reference grid made of lines that intersect to reference two or three dimensional locations or absolute locations. The first type of coordinate systems examined, in regards to GIS, are geographic coordinate systems. Absolute location may be used on two dimensional or three dimensional coordinate systems. Longitude and latitude are used in measuring absolute location on the earth (a three dimensional sphere) with geographic coordinate systems. This **geographic coordinate system** uses two sets of imaginary lines around the earth, **parallels**, representing latitude, and **meridians**, representing longitude, forming a "grid" over the earth.

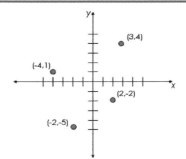

Coordinate systems, as shown in this example, are used to find locations. Geographic coordinate systems use similar methods to find absolute locations on the earth's surface.

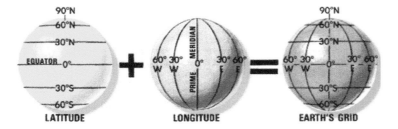

LATITUDE   LONGITUDE   EARTH'S GRID

Latitude

Lines of **latitude** run east and west along the earth's surface never intersecting and are also called **parallels**. The starting point for the lines of latitude is at the horizontal center of the earth, the **equator**. These lines represent a measure of the angle formed from the equator to the specified point north or south. The earth is a sphere with 360° which is measurable with any size globe or spherical representation of the earth. Using angles to measure location provides an easy, universal method for measurement. An easy way to remember latitude and its location in the coordinate system is that the parallel lines of latitude resemble the rungs on a ladder.

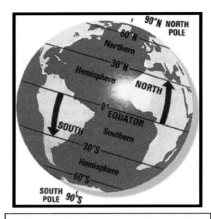

Lines of Latitude run east to west. The centermost line is the equator. http://www.hammondmap.com/catal og/classroom_activities/latlong1.html

The prime parallel or line of latitude is the equator. The equator is an imaginary line that divides the Earth into the northern and southern hemispheres. The equator is measured as 0° latitude with 90° separating it from each pole. Latitude lines are measured 0° to 90° north to the North Pole and 0° to 90° south to the South Pole.

On a globe of the Earth, lines of latitude are circles of descending sizes from the equator to the poles. The longest is the equator, whose latitude is zero, while at the poles - at latitudes 90° north and 90° south - the circles shrink to a point.

In addition to the equator, there are two other significant imaginary latitudinal lines above and below the equator; the **Tropic of Cancer** and the **Tropic of Capricorn**. The Tropic of Cancer divides the tropics from the northern temperature zone; located at 23°30' north latitude. The Tropic of Capricorn divides the tropics from the southern

The area in light blue is the Tropics. The year round temperature in this area does not fluctuate very much at all. Image source: http://www.worldatlas.com/aatlas/imagee.htm

temperature zone; located at 23°30' south latitude. The area, or zone, of the earth between the two lines is known as the "**Tropics**." There are no dramatic changes in seasons within this zone because the sun is always directly overhead. People living north of the Tropic of Cancer and south of the Tropic of Capricorn however do experience dramatic seasonal climate changes. Based on the angle of the sun, when it's summer north of the Tropic of Cancer, it is winter south of the Tropic of Capricorn.

Longitude
Lines of **longitude** that run north and south along the earth's surface are also called **meridians** and intersect at the poles. You can remember these by the word "long" and longitude, and longitude lines span the length between the two poles. The word meridian is derived from the Latin, meri, a variation of "medius" which denotes "middle", and diem, meaning "day." The word meridian is synonymous with "noon".   All points on the same line of longitude experienced noon (and any other hour) at the same time and were therefore said to be on the same "meridian line", which became "meridian" for short. The Sun at noon was said to be "passing meridian". Before noon is known as "ante meridian", while times after it are "post meridian." Today we use the abbreviations a.m. and p.m.

Unlike latitude lines, meridians do intersect at the poles of the earth. The **prime meridian,** measured as 0° longitude, is the starting point for the meridians. The lines of longitude are measured from 0° to 180° west and 0° to 180° east of the prime meridian. Those located to the west of the prime meridian represent locations in the **western hemisphere** and those east of the prime meridian represent locations in the **eastern hemisphere**. Another important line of longitude is the **International Date Line** and is located roughly 180° from the prime meridian. The International Date Line is the imaginary line that separates two calendar days. If you look on a map or globe, you will see the International Date Line is not a straight line; it is established at 180° but shifts around any countries located on the date line to avoid one country operating in two calendar days.

The prime meridian runs through Greenwich, England. Historically, the meridian passing through the old Royal Astronomical Observatory in Greenwich, England, was chosen as zero longitude. Located at the eastern edge of London, the British capital, the observatory is now a public museum and a brass band stretching across its yard marks the "prime meridian." Tourists often get photographed as they straddle it--one foot in the eastern hemisphere of the Earth, the other in the western hemisphere.

Reading Latitude from a Map or Globe
Remember that lines of latitude (parallels) run east to west measuring location north or south of the equator.

To read latitude from a map:
• Find the equator and determine if the location is north or south of the equator.
• Find which two lines of latitude the location lies between.
• Find a midpoint between the lines of latitude. Determine if the location is closer to the midpoint or one of the lines of latitude.
• Estimate the latitude of the location in degrees.

When estimating latitude for a location, it is feasible to estimate degrees only. It is difficult to estimate minutes and seconds with the eye only. It is important to understand how to estimate absolute location with a map. However, there are numerous websites that provide absolute locations for almost every city in the world. In a GIS latitude (and longitude) is measured continuously as you navigate the map display, but the printed layouts require knowledge of reading and estimating coordinate systems.

Directional coordinates for latitude are ALWAYS north or south – NEVER east or west.

Reading Longitude from a Map or Globe
Lines of longitude (meridians) run north to south measuring location east or west of the Prime Meridian.

To read longitude from a map:
• Find the Prime Meridian and determine if the location is east or west of the Prime Meridian.
• Find which two lines of longitude the location lies between.
• Find a midpoint between the lines of longitude. Determine if the location is closer to the midpoint or one of the lines of longitude.
• Estimate the longitude of the location in degrees.

Notice that the steps you need to take in reading the longitude of a location are almost identical to those for reading the latitude. The only real difference is the direction you are looking.

Directional coordinates for longitude are ALWAYS east or west – NEVER north or south.

Expressing Absolute Location in Geographic Coordinates
When reading absolute location from a world map or globe, it is suitable to estimate to the nearest degree, but to report a more accurate absolute location for a more specific location or for reading a map with a higher degree of accuracy, it is necessary to divide the distance between degrees further. One way to do this is to use decimals to represent coordinates more accurately, called Decimal Degrees. A second method for representing degrees divides each degree into sixtieths called minutes and further divides minutes into sixtieths called seconds – similar to the way an hour is also divided into sixtieths, called Degrees Minutes Seconds. In this method, thirty minutes is half of sixty and therefore is half way between two degrees.

Absolute location expressed:
In Degrees Minutes Seconds (with direction)
Ex: 30o15'30" W, 45 27'45" N
In Decimals Degrees (with negative (-) for West longitudes and South latitudes)
Ex: -30.2583, 45.4625

In some cases depending on preference, it may be necessary to translate between the two formats. We will discuss and use the formulas to convert from Degrees Minutes Seconds (DMS) to Decimal Degrees (DD) and vice versa. Understanding this process is essential to understanding geographic coordinate systems in general. However, there are many automatic converters that are located on websites that automatically perform the conversion for you should you need them. You simply enter the DMS or DD and the converter gives you the coordinate in the other format. Similarly, ArcGIS allows users to quickly change measurement display formats.

Longitude Latitude Conversion
The examples below are of a longitude conversion from Degrees Minutes Seconds (DMS) to Decimal Degrees (DD) and from DD to DMS. These same conversion formulas work for latitude coordinates as well. To convert your own coordinates from one format to another, replace the numbers in either example with the coordinate values of your location.

Notice that a negative sign (-) is added to the DD coordinate because this coordinate is west of the Prime Meridian. DD longitude coordinates west of the Prime Meridian and DD latitude coordinates south of the Equator are expressed as negative numbers. You do not include a negative sign (-) when using DMS coordinates – you use cardinal directions (N,S,E,W) instead.

DMS to DD	DD to DMS
122º 10' 9" W 9/60 = 0.15 minutes 10 + 0.15 = 10.15 minutes 10.15 / 60 = 0.1691 122 + 0.1691 = 122.1691 **-122.1691 DD Longitude**	**-122.1691 DD Longitude** 122.1691-122 = 0.1691 0.1691 x 60 = 10.15 = 10 minutes 0.15 x 60 = 9 = 9 seconds **122º 10' 9" W**

A feature's absolute location may be given in Degrees Minutes Seconds format, OR in Decimal Degrees format. Little Rock, Arkansas is located at 35° 13' 12" 92° 22' 4"W or 35.22, -92.3678.

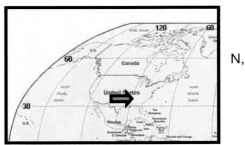

N,

Relative Location
Relative location is describing where an object is on the earth's surface without giving absolute coordinates. This method can be used when a map or globe is unavailable to describe location. Relative locations are used by describing how far a feature is from other features. Describing how far your house is from school or how far it is from your work to the beach are examples of relative location.

Relative location is an approximation. To say that your cousin's house is only 30 minutes from the beach probably means that it takes about 30 minutes to drive to the beach from their house. It might actually only be 15 miles but takes 30 minutes due to traffic conditions and speed limits. A relative location description can change based on the features you use to compare. For example, Little Rock is located approximately 50 miles north-northwest from Pine Bluff, Arkansas and it is also located 100 miles southeast of Clarksville, Arkansas. This method is most effective when the person receiving the location is either in the area or familiar with the area.

Conclusion
A geographic coordinate system is used to determine a location on a globe or sphere using grid system to establish absolute location. This grid system uses parallels of latitude and meridians of longitude to measure an object's location based on its distance north-to-south to either pole from the Equator; and east-to-west from the Prime Meridian.
Absolute location may be represented as *x* and *y* coordinates expressed as Degrees Minutes Seconds (DMS), or Decimal Degrees (DD).

Relative location is used to determine how close or far, and in what direction a landscape feature is from other landscape features. This method can be used when a globe or map is not available.

Technology provides ways to quickly measure absolute location and knowledge of these systems will greatly enhance your knowledge for GIS.

Lesson 2: Absolute Location Exercise

Directions: This exercise practices reading the absolute location of a particular place on a map using longitude and latitude coordinates. In addition, you will convert absolute location coordinate formats. Access a world map and atlas and the notes taken from the PowerPoint presentation to answer these questions.

Absolute Location:
What is the approximate absolute location (in degrees) of the following cities?

City Name	E or W	Longitude	N or S	Latitude
Example: Washington, DC	*W*	*76*	*N*	*39*
1. San Diego, CA				
2. Cape Town, S Africa				
3. Sydney, Australia				
4. London, England				
5. New Orleans, LA				
6. Moscow, Russia				
7. Miami, FL				
8. Paris, France				

Converting DMS to DD:
Part I. Convert the following absolute locations from Degrees Minutes Seconds (DMS) to Decimal Degrees (DD). Show your work.
Part II. Name the location of these coordinates on the world map.

Example: 165°35' 22" E

22/60 = 0.366667 minutes
35 + 0.366667 = 35.366667
35.366667/60 = 0.589444
165 + 0.589444 = 165.289444
-165.289444   DD Longitude

75°3' 58" S

58/60 = 0.966666 minutes
3 + 0.966666 = 3.966666 minutes
3.966666/60 = 0.066111
75 + 0.066111 = 75.066111
75.066111 DD Latitude

Location: East Siberian Sea

6.   20°47' 5" W   89° 17' 45" N | Location: |

7.   10° 56' 32" E   55° 21' 13" N | Location: |

Converting DD to DMS:

Part I. Convert the following absolute locations from Decimal Degrees to Degrees Minutes
Seconds. Show your work.

Part II. Name the location of these coordinates on the world map.

Example: -122.1681 Longitude   *37.8514 Latitude* | Location: San Francisco, CA |

122.1682 – 122 = 0.1681   *37.8514 – 37 = 0.8514*
0.1681 x 60 = 10.086 = 10 minutes   *0.8514 x 60 = 51.084 = 51 minutes*
0.086 x 60 = 5.16 = 5 seconds   *0.084 x 60 = 5.04 = 5 seconds*
122° 10' 5" W   ***37° 51' 5" N***

8.   - 56.7631 Longitude   54.6732 Latitude | Location: |

9.   108.8765 Longitude   - 43.0845 Latitude | Location: |

Application Questions
Directions: Answer the following questions using the information from the lesson or the notes from the PowerPoint presentation. In instances where two possible answers are given, circle the correct answer.

10. A longitude measurement taken west of the prime meridian is going to be a (**positive** or **negative**) number when displayed in decimal format.

11. Using DMS method of stating absolute location for longitude, only _____ and _____ are used to determine the direction from the prime meridian.

12. Lines of latitude are measured from _____ to _____ degrees.

13. Lines of longitude are measured from _____ to _____ degrees.

14. The _____ _____ _____ is located roughly 180° from the prime meridian.

15. Lines of latitude never touch. (**True** or **False**)

16. Directional coordinates for latitude are always east or west. (**True** or **False**)

17. Describing where an object is on the earth's surface without giving absolute coordinates is giving its _____ location.

18. A geographic coordinate system is a way to measure angular distance on a sphere. (**True** or **False**)

Lesson 2: Introduction to Coordinate Systems and Location Lesson Review

Key Terms
Use the lesson or glossary provided in the back of the book to define each of the following terms.

1. Absolute location
2. Eastern Hemisphere
3. Equator
4. Geographic coordinate system
5. International Date Line
6. Latitude
7. Longitude
8. Meridians

9. Parallels
10. Prime meridian
11. Relative location
12. Tropic of Cancer
13. Tropic of Capricorn
14. Tropics
15. Western Hemisphere

Global Concepts
Use the information from the lesson to answer the following questions. Use complete sentences for your answers.

16. What is the highest degree of measurement that can be stated for latitude? Would this measurement be located at the equator or one of the poles? Explain your answer.

17. What are the names of the imaginary lines that border the Tropics on both the north and south ends? Are there drastic changes in the weather in the Tropics? Explain you answer.

18. What is the highest degree of measurement that can be stated for longitude? What is the name for the imaginary line that is the starting point for lines of longitude?

19.   What does the International Date Line signify? Why is it not a straight line?

20.   The steps for reading Latitude and Longitude on a map or globe are very similar. Write the steps for finding Latitude below.

Let's Talk About It...
Answer the following question and share the responses with your instructor and classmates.

21.   Discuss how relative location is different than absolute location. Why would one method be used over another?

22.   Compare Degrees Minutes Seconds versus Decimal Degrees. How are they similar? How are they different?

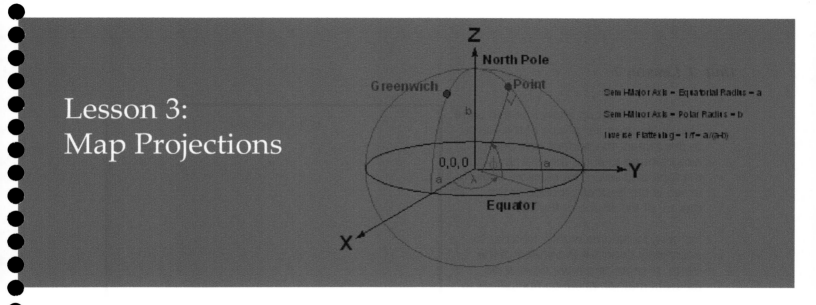

Lesson 3:
Map Projections

Picture a map of the earth presented to you on the skin of an orange. You are given the task of taking off the peel and to flatten it out thus taking a three dimensional object and turn it into a two dimensional object. Have you ever taken an orange peel off of an orange and tried to flatten it out? It is impossible to do without stretching or tearing the peel. Since the earth that we live on is not flat but is spherical, it is a constant challenge to create a two dimensional map that most accurately represents the three dimensional earth. Projected coordinate systems allow for a more accurate translation of the actual shape of the earth to a two dimensional surface. In the previous lesson, geographic coordinate systems were used to reference location on the spherical earth. This lesson will focus on two processes that allow users to display features more accurately in a GIS. First to improve accuracy is the concept that the earth is not a perfect sphere. In order to accurately state where something is located on the surface of our planet, the true shape of it must be taken into account. Once the earth's shape is accounted for, then the focus shifts to representing that shape in a two dimensional environment that is used by maps and the ArcMap display.

Shape of the Earth

Historically, the best representation of the Earth has been the globe. Globes may come in different colors and sizes, some with texture, some smooth, but they all have one thing in common, they are based on a perfect sphere. Although globes provide a decent replica, this is not exactly a true representation of the shape of our planet. The Earth is slightly flattened at the poles, making it an **ellipsoid** rather than a perfect sphere. In order to make accurate distance calculations and estimates of the earth's area and volume, the earth must be modeled more accurately. How can this be done? Through the use of models called geodetic datums.

Historically, globes have been used to represent the shape of the Earth.

Geodetic Datums

There are many different models made to represent the shape of the Earth. According to the ESRI A to Z GIS dictionary of terms, a **geodetic datum** is "the basis for calculating positions on the earth's surface or heights above or below the earth's surface".

The example datum shown on this page shows the irregular shape of the Earth. In this image, the surface has been exaggerated to show the elevation of the land and sea. The ellipsoid shown in red is the geodetic model, or the datum, of the shape of the Earth.

So how do datums work? In order to accurately describe the shape of the Earth, researchers have developed different mathematical computations to serve as models of the earth. Researchers who develop geodetic datums build mathematical equations based on their best estimation of the geometry of the Earth to model the

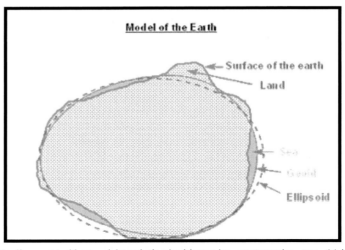

http://www.og.dti.gov.uk/regulation/guidance/co_systems/co_sys_11.htm

Earth's three dimensional shape using reference points around the earth. Often in GIS and with GPS units, the datums that are used calculate positions on the earth's surface and are called **horizontal geodetic datums**. Other datums as noted in the definition account for heights, and these are called **vertical geodetic datums.** As GIS users, it is not necessarily imperative to understand the mathematics used to create the datum or to memorize all of the datums. It is, however, important to understand that different datums will describe the earth in different ways. For instance, GPS units have settings to specify which datum is used in gathering information. The GIS user needs to know this information before uploading that data to ArcMap. If different datums are used, ArcMap may have a conflict in displaying data accurately. Let's look at a few datums more closely.

One of the earliest datums was calculated in 1927 and was based on the Clarke 1866 ellipsoid. It was calculated using traditional survey and computation equipment of the era, such as telescopes, pendulums, gravity meters, and logarithmic scales. Because the shape of the Earth is an ellipsoid, the polar radius is slightly less than the equatorial radius.

Clarke 1866 (NAD27) and GRS1980 (NAD83) depict different models of the earth and represent the change in describing the earth as new technology develops

When satellites and more accurate computation and surveying tools became available, a more accurate datum was developed. The North American Datum of 1983 (NAD83) uses a slightly different set of geometric parameters to define the shape of the Earth, and hence enables more accurate distance measurements.

The differences between NAD27 and NAD83 are minimal; however, when compared with the size of the earth, though they can account for a difference of up to 100 meters in some areas of variable land masses. Again, as GIS users, it is important to be familiar with the datum to be used in a GIS because 100 meters is a significant amount when precision is necessary.

This chart shows a sample list of the many datums that are commonly used. The equatorial and polar radii are provided, as well as the flattening (the difference between the two radii estimates). This chart also provides the most common uses for each datum. It is important to remember that none of these perfectly describe the earth's shape nor should one be thought of necessarily as better. Some datums are preferred by some organizations as noted on the graphic, and some datums represent certain areas more accurately. Once a datum is specified, users are able to more accurately represent location in a GIS.

Datum	Equatorial Radius, meters (a)	Polar Radius, meters (b)	Flattening (a-b)/a	Use
NAD83/WGS84	6,378,137	6,356,752.31	1/297.257223563	Global
GRS 80	6,378,137	6,356,752.31	1/297.257222101	US
WGS72	6,378,135	6,356,750.50	1/298.26	NASA, DOD
Australian 1965	6,378,160	6,356,774.70	1/298.25	Australia
Krasovsky 1940	6,378,245	6,356,863.00	1/298.3	Soviet Union
International (1924) -Hayford (1909)	6,378,388	6,356,911.90	1/297	Global except as listed
Clake 1880	6,378,249.10	6,356,514.90	1/293.46	France, Africa
Clarke 1866	6,378,206.40	6,356,583.80	1/294.98	North America
Airy 1830	6,377,563.40	6,356,256.90	1/299.32	Great Britain
Bessel 1841	6,377,397.20	6,356,079.00	1/299.15	Central Europe, Chile, Indonesia
Everest 1830	6,377,276.30	6,356,075.40	1/300.80	South Asia

Hundreds of datums have been created to describe the irregular shape of our earth.

Geographic coordinate systems described in Lesson 2 provide reference for location on a sphere, such as the Earth. Datums describe the Earth and its imperfections more accurately but are still based on a spherical shape; therefore, when a datum is applied, geographic coordinate systems are still relevant to reference a specific location. This means that longitude and latitude still apply when a datum is specified.

Map Projections
Mapping and GIS both strive to create accurate depictions of the earth. With datums, accuracy was increased by describing the earth beyond that of a simple sphere. Now that the earth's shape is described more accurately, that shape can be projected onto a two dimensional surface. The best way to illustrate the difficulty in doing this is by thinking of the Earth as a rubber ball with the land and water painted on it. To flatten the rubber ball into a flat surface we need to cut it up and stretch it. Because the rubber ball is being stretched, the land shown on it will be distorted from its original shape. This same cutting and stretching must be performed on the earth to make a map by using a mathematical formulae. This process results in a **map projection**. Projection formulae take the geographic coordinates from the spherical earth (longitude and latitude; and possibly with a datum applied) and convert them to projected coordinates (also displayed in *x,y* format like longitude and latitude).

Whether the rubber ball is actually being flattened or the earth is being theoretically or digitally flattened, the tearing and stretching is still a factor. In both cases

http://www.ga.gov.au/nmd/geodesy/datums

distortion occurs. Distortion cannot be eliminated completely, but the goal of a projection is to minimize that distortion. There are four properties of the Earth's surface that will be distorted and are important to consider when using a map projection.

---

1. **Area** - On a world map, for example, you may want to preserve the relative area of different countries so that a distribution of some type of geographic phenomenon may be shown.

2. **Shape** – For educational purposes, it may be important for a wall map to preserve the shapes of states or countries so that students can learn to identify them.

3 & 4. **Direction & Distance** - The most important properties for navigational purposes.

Because there is not one specific projection that can accurately preserve all four of these properties: area, shape, direction, and distance; there will always be distortion. It is best to use a projection that preserves the property most useful to the study.

Azimuthal Equidistant Projection

Some map projections are planar, or azimuthal, in nature. In other words, these projections are created as if a flat piece of paper (a plane) is placed touching the globe and the geographic features on the globe are projected onto the paper. Distortion would be minimal for the area that is closest to the place that the paper touches but would increase drastically as you move from that point. These types of projections are typically useful for illustrating a hemisphere. The example given to the right shows the Polar Azimuthal Equidistant projection. This projection allows all distances from the center of the map along any of the longitudinal lines to be accurate, thus distance is the property most accurately preserved. As with all projections, as you travel away from the center of the projection, distortions occur.

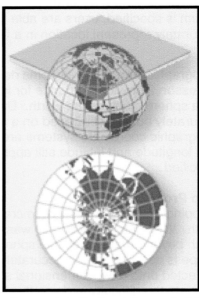

Keep in mind that projections are mathematical equations that are used based on the geometry of the type of map projection to convert the 3D coordinates to 2D coordinates.

Source: http://nationalatlas.gov/articles /mapping/a_projections.html

Conic Projection

A conic projection is computed as if a piece of paper is shaped as a cone and placed around the globe (like an ice cream cone). In the area that "touches" the paper, the distortion is less than the other areas where the paper does not touch. The example given below shows the Lambert Conformal Conic projection, which is best suited to

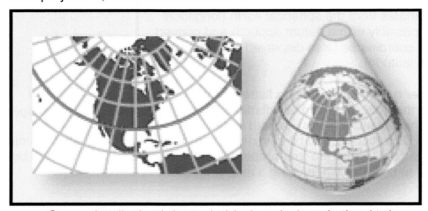

Source: http://nationalatlas.gov/articles/mapping/a_projections.html

---

mapping mid-latitude regions in either the northern or southern hemispheres. The conic projection most accurately displays area.

Cylindrical Map Projections

Cylindrical map projections simulate a piece of paper wrapped around the globe. Some of these projections mimic the paper touching the globe at the Equator. Others simulate a transverse orientation, meaning the map's orientation is based off of a line of longitude like the Prime Meridian rather than a line of latitude like the Equator. The example given shows the Mercator projection, which is generally used for maps of the entire Earth, since they tend to avoid the extreme distortion that occurs using other projections.

Conclusion

As cartographers learned more about the Earth, they realized that its shape was not a perfect sphere. To allow for variances in the

Source: http://nationalatlas.gov/articles/mapping/ a projections.html

earth's surface, mathematical models were created to simulate the Earth's shape. These models are called geodetic datums.

The most accurate reproduction of the Earth is the globe, since both are relatively spherical in shape. However, map projection techniques are used by cartographers to represent the 3D earth more accurately on a flat 2D surface. The goal of a map projection is, with as much accuracy as possible, to represent the area, shape, direction, and distance of the Earth's land and water masses. Each map projection retains one property of the Earth better than others and has its own advantages, and its own limitations. In using a map projection for a project, a map projection should be chosen based on which projection most accurately represents the geographic data being studied.

For GIS users, knowledge of datums and projections can increase the accuracy of analysis and validity of maps by more accurately representing the Earth. Rather than performing analyses and making observations based on a perfect sphere, realize, first, that the Earth is not a uniform sphere but rather an ellipsoid requiring a datum. Second, use projections to more accurately represent the Earth's shape in a two dimensional environment.

Lesson 3: Map Projections Exercise

This lesson and presentation have provided you with information about datums and map projections. Sometimes, though, it is easier to understand a concept if you see it in practice.

Directions: Use this lesson, your notes from the presentation, the USGS Map Projections Poster or website, and observations from the map projection exercises to answer the following questions.

1. Why are globes not entirely accurate in depicting the true shape of the Earth?

2. As a GIS user, explain why it is important to know what datum your GPS unit is in when collecting points?

3. How are the two North American datums (NAD27 & NAD83) similar? How are they different?

4. Why would one datum be more beneficial to use over another datum?

5. Why are map projections used? What is the goal of a projection?

6. Name three different projection types and describe their characteristics and appropriate uses.

7.   In the "rubber ball" example under Map Projections, when the ball was flattened, what steps could be taken to minimize distortion?

8.   Can distortion be eliminated completely? Explain.

To answer the following questions use the **USGS Map Projections Poster**.
(Note that the USGS Map Projections Poster is also available online at http://erg.usgs.gov/isb/pubs/MapProjections/projections.html)

What are the type, areas of least distortion, and primary uses for the following projections?

Projection	Projection Type	Least Distorted Areas	Primary Uses
5. **Mercator**			
6. **Albers Equal Area Conic**			
7. **Orthographic**			
8. **Stereographic**			

Optional Exercises –

Map Projections Demonstration:
To illustrate a map projection by converting a three-dimensional globe to a two-dimensional surface:

1. Inflate the inflatable globe, if you haven't done so already.

2. Note the location of important meridians and parallels, as well as continents.

3. Using whatever means necessary, make the three-dimensional globe lay flat – that is, create a map projection from an inflatable globe. You may use scissors supplied by your instructor. Refer to the **USGS Map Projections Poster** or the link at www.SPACESTARS.org for ideas on how to do this.

4. Once you have finished, lay the two-dimensional globe on a flat surface and compare the two-dimensional globe with a three-dimensional globe.

5. Compare your projection with the projections on the poster.

6. Briefly describe to your classmates how you created your map projection and why you chose the method that you did.

Important Points to Note and Discuss:

- What parts of the projected globe have changed? What caused the change?

- What parts did not change? Why?

- Which projection on the **USGS Map Projections Poster** most closely matches the projection that you just created?

- What other method could have been used to flatten the globe instead of cutting the globe numerous times?

Lesson 3: Map Projections Lesson Review

Key Terms
Use the lesson or glossary provided in the back of the book to define each of the following terms.

1. Ellipsoid

2. Geodetic Datum

3. Horizontal geodetic datum

4. Vertical geodetic datum

5. Map Projection

6. Azimuthal Projection

7. Conic Projection

8. Cylindrical Projection

Global Concepts
Use the information from the lesson to answer the following questions. Use complete sentences for your answers.

9. Globes have been used as miniature models of the Earth historically. Explain why they do not provide a true representation of the Earth's surface.

10. Is there one datum that is considered better than the others? Why would one datum be preferred over another datum?

11.   Identify the four properties of the Earth's surface that will be distorted when using a map projection.

12.   When was the original datum calculated and how is it different than NAD83?

Let's Talk About It...
Answer the following question and share the responses with your instructor and classmates.

13.   Of the projections listed in this lesson, which would you prefer to use if your study are included land mass at the equator? Why?

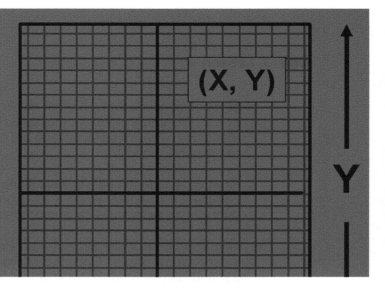

Lesson 4:
Map Coordinate Systems

In Lesson 2, you learned how to determine the absolute location of a feature on the surface of the Earth, closely resembling a three dimensional sphere, using the Geographic Coordinate System. Because it was designed specifically for a three dimensional sphere or spheroid, it uses angular measurements consisting of degrees, and can also include minutes and seconds to be more precise. In Lesson 3, the fact that the earth is not perfectly round became a factor in describing the Earth's true, three dimensional shape using a datum. Now that the Earth's representation is more accurate due to datums, we can transform that shape, or model, into a two dimensional surface to be displayed on a map. These transformations are accomplished through projecting the model. The previous lessons have documented the process to take the Earth from a 3D sphere to 3D spheroid (via datums) to a 2D map (via projections). When identifying location on the surface of a sphere, or globe, absolute location was found using the geographic coordinate system, but now the earth is represented in a map, which gives us more options in determining absolute location. In this lesson, you will look further at coordinate systems and how they are represented on a projected, or two dimensional, map surface. The three most commonly used coordinate systems that GIS analysts may see are the **Geographic Coordinate System**, **Universal Transverse Mercator** (**UTM**) and **State Plane**.

Coordinate Systems
So what are coordinate systems again? In Lesson 2 of this unit, you were most likely reintroduced to the concept of coordinate systems. Just like in your math class, on a grid, you can plot points based on their X and Y coordinates.

Coordinate systems in geospatial technology follow the same procedures as coordinate systems in math. Geospatial coordinate systems are also called spatial reference systems or coordinate reference systems. These are all terms for a way to communicate where something is located on the surface of the Earth using an x,y coordinate.

Coordinate systems are used to describe locations using x,y coordinates

Geographic Coordinate System
The Geographic Coordinate System uses the longitude and latitude grid that is draped over the Earth's surface. In the Geographic Coordinate System, the longitude coordinate provides the X value, and the latitude coordinate provides the Y value. Therefore, a point on the grid is represented as (Long, Lat). To recap Lesson 2:

- Lines of longitude that run north and south along the earth's surface and meet at the poles. The prime meridian, measured as 0° longitude, is the starting point for the meridians. The lines of longitude are measured from 0° to 180° west and 0° to 180° east of the prime meridian.

- Lines of latitude run east to west along the Earth's surface never intersecting. The prime parallel is the equator which is measured at 0° latitude. Latitude lines are measured 0° to 90° north to the North Pole and 0° to 90° south to the South Pole.

The Geographic Coordinate System was primarily designed to represent the angular distance of a three dimensional object. Once this coordinate system is projected onto a flat surface, a longitude and latitude reading is a way to reference a point on the Earth as a sphere. It is not, however, the best way to measure between two points on the earth. To say that: "Today, I am going to travel 10° south", does not exactly give the average person any idea of how far or where exactly you are traveling. On the other hand, because the projected GCS is good for referencing points on the three dimensional Earth, it is widely used when hurricanes, cyclones, or other global events are plotted.

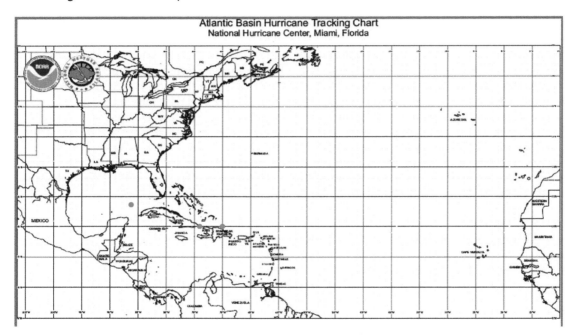

Once a longitude and latitude coordinate is given, the coordinate can be located on the map and a mark can be made designating the current position of the eye of the storm. New marks can be made as new coordinates are given.

This is a reduced version of the chart used to track hurricanes at the National Hurricane Center.
http://www.nhc.noaa.gov/HAW2/pdf/AT_Track_chart.pdf

UTM

The National Imagery and Mapping Agency (NIMA) adopted a special grid, or coordinate system, to be used for military purposes called the Universal Transverse Mercator (UTM) grid. Unlike the GCS, this grid is not based on a global or politically divided projection. In this grid, the world is divided into 60 zones, with each zone covering a strip 6° wide in longitude. The starting point, or Zone 1, for UTM is approximately the International Date Line goes east through Zone 60. The contiguous United States are contained within UTM zones 10 to 19. By "slicing" the Earth into zones, each zone is shown with greater accuracy. Recall the example in

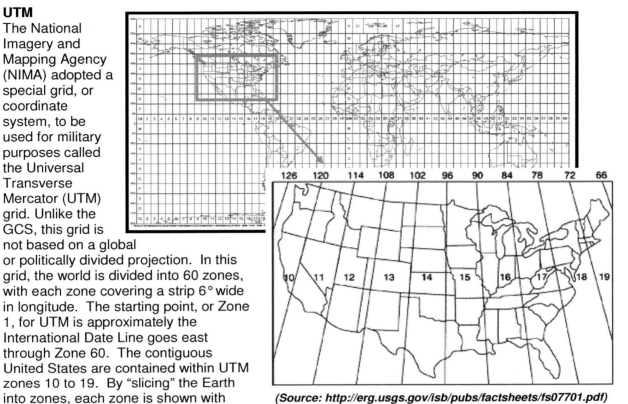

(Source: http://erg.usgs.gov/isb/pubs/factsheets/fs07701.pdf)

the previous lesson of the orange peel that was flattened. Imagine that you are flattening half of the peel (or a hemisphere) by placing it on a table and pressing down with your thumb until it flattens. This is a difficult task, but what if we cut off a small section of the orange and did the same the process. Flattening the smaller section would be easier and with a small enough piece, you might not notice distortion at all - the same concept applies to the projection in UTM Coordinates. Many NIMA and USGS-produced maps rely on the UTM coordinate system because of the lessened distortion.

So how is this coordinate system unique? The projected zones are now "flat", which means we can use a two dimensional coordinate system. Within each UTM zone, coordinates are measured east and north in meters. The UTM system reports the X coordinate, the **Easting** value, and the Y value, the Northing Value, in meters. Ultimately, you are going to figure out how many meters you are from the western most edge of the zone and how far north you are from the Equator. The easting values begin at 0 meters from the western edge to 1,000,000 meters on the eastern edge, so that every measurement given tells us how far east of the western boundary we are located. The northing values are measured from zero at the Equator to 10,000,000 meters at the North Pole. For locations south of the Equator, cartographers assigned the Equator a false northing value of 10,000,000 meters to avoid negative numbers extending to zero at the South Pole. A coordinate system measure in meters allows for quick determination of distance and, also, allows for calculations like.

State Plane
A third coordinate system that a GIS user might use is the State Plane system. The State Plane coordinate system is a system that is used exclusively in the United States.
Just as the UTM coordinate system focuses in on a "strip" of the Earth thus making it more reliable than larger projections, the State Plane focuses in on yet a smaller piece of the earth, a single state or a smaller portion of a state. State Plane is popular among state and local governments due to its high degree of accuracy and the United States' focus on English measurements (yards, feet, inches, etc.). State Plane is not as useful for GIS projects, however, when the extent of a project extends beyond the boundaries of a particular zone. Each zone is calculated for use only with locations that fall within the zone. If your project uses location outside of a particular zone, then you should select a coordinate system that is intended for a larger area, like UTM.

For the State Plane system, coordinates are given in feet with the Easting value providing the X value and the Northing value provided the Y value. With this coordinate system, just like with the UTM coordinate system, it is easy to determine distances between two features, especially when they are within the exact same zones of the state plane.

http://www.vterrain.org/Projections/spcs.html

Conclusion
Once our three dimensional Earth is projected on a two dimensional surface, we gain the ability to use planar coordinate systems. Using   coordinate system that is best suited for your needs is a valid concern for any GIS user. Coordinate systems are used to define points on the Earth using Cartesian (x,y) coordinates. Three most commonly used coordinate systems in the US are the geographic coordinate system (long/lat), UTM and the State Plane Coordinate System. Each has their own pros and cons. The Geographic Coordinate System is useful when describing locations with respect to there location especially in projects that have a global focus. UTM and State Plane systems are useful when focusing on a smaller extent and for showing distance. When using GIS, it is important that the project manager takes into consideration the size of the study area before deciding which coordinate system is best to use and the audience when planning begins. Is the map intended for use by federal agencies which rely often on UTM. Is it an international project with audiences in countries that are accustomed to the metric system? Is it In order for data to align properly, the same datum, map projection and coordinate systems must be used for all data sets in the same project.

Lesson 4: Map Coordinate Systems Exercise: Reading UTM Maps

Maps are projected using different mathematical algorithms or formulas to reduce the amount of distortion that occurs when viewing portions of the round earth on a flat screen or paper. When viewing flat maps using the Geographic Coordinate System with longitude and latitude, or viewing local, smaller areas (like your school), these maps experience little enough distortion to not significantly affect the shape or distance of landscape features. When studying regional areas at small scale (like your city or county), this distortion is enough to warrant correction, or *rectification*, of the error. The projection most commonly used by the United States Geological Survey (USGS) to rectify distortion is the Transverse Mercator (TM) Projection. This projection uses the vertical center line in each zone as its center point (as opposed to using the equatorial or horizontal center point). The TM projection, coupled with a systematic metric grid coordinate system, combine to make up a worldwide system of UTM projection zones used by the USGS to map the full extent of the United States.

The United States is divided into 10 UTM zones, each measuring 6° of longitude wide. *(Figure 1)*

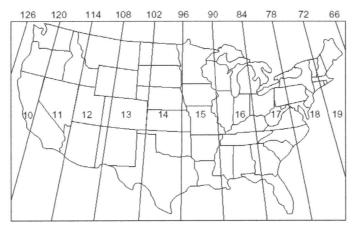

Figure 1 UTM Zones
(Source: http://erg.usgs.gov/isb/pubs/factsheets/fs07701.pdf)

Each zone is measured in meters from west to east (known as **easting**), and from south to north (known as **northing**). Ultimately, UTM coordinates are providing information to allow the reader know both how many meters east of the International Date Line and how many meters north of the equator a location is. An arbitrary false northing value of 10,000,000 meters is placed at the equator to avoid negative coordinate values. However, for locations north of the equator, distance from the equator is measured as meters starting from 0. Locations south of the equator use the 10,000,000-meter equatorial designation and are measured from the equator starting at 10,000,000 m.

Reading the UTM Grid on a USGS Topo Map:

To ensure that your map includes the information that is important to your study area, first verify the zone number. Because the location is within the contiguous United States, this number will correspond with one on the UTM zone map shown on the previous page. The UTM zone is identified in the lower left corner of the map:

In this example, the UTM zone is 15.

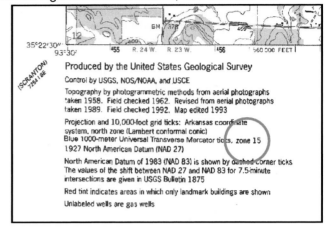

Each USGS topographic map has a series of blue tick marks placed to mark every 1,000 meters on the ground along all four sides of the map sheet,

or…

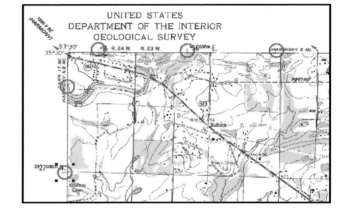

a series of vertical gridlines superimposed on the map to mark the 1,000 meters .

Each of the cells in between these marks represents 1000 square meters.

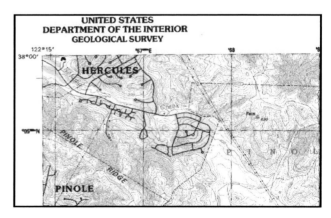

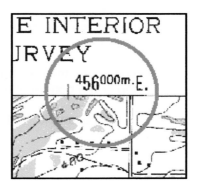

Beside the first or second tick mark or gridline in the northwest corner of the map is the **x** or **easting value** of that location. In this example, any landscape feature that corresponds to this blue tick mark identified along the top and bottom of the map sheet is 456,000 meters east of the UTM zone boundary.

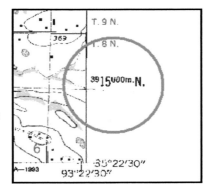

Likewise, a similar set of blue tick marks or horizontal gridlines and designations are made along the sides of the map sheet to locate the **y** or **northing value** of a landscape feature. In this example, any landscape feature that corresponds to this tick mark identified along side of the map sheet is 3,915,000 meters north of the equator.

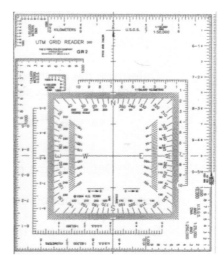

For locations within the perimeter of the gridlines is made easier with the use of a UTM grid reader. These are also referred to as UTM Corner Rulers because you can use the various corners or elbows to determine positions on the map. Once the transparent grid reader is placed on top of the map, with the corner at the destination, more precise UTM x,y coordinates can be determined.

Determining the UTM *XY* Coordinates for Landscape Features on a USGS Map Sheet

Use the following instructions to determine the UTM coordinates for features in your local community:

1. Determine the scale of your community map. This information will be located on the bottom area, just below your map. An example: SCALE 1:24,000 METERS

2. Find the same scale representation on your UTM grid reader.
3. Locate the feature on the map. This indicates a campsite (for this example, look for the ★).

4. Determine the easting grid value. This will be the number on the tick mark at either the top or the bottom of the map and is the one that is to the left of the feature.

    Example: ⁷61 At this point, you already know that the object is at least 761,000 meters east of the UTM zone boundary.

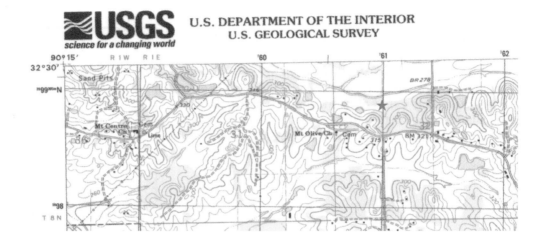

5. Determine the northing grid value next. This will be the tick mark or grid line below the feature. It will be located on both the left and right margins of the map.

    Example: ³⁵98 You know now that the feature is located at least 3,590,000 meters north of the Equator.

    Determining a more exact coordinate will require the UTM Corner Grid Reader.

6. Locate the corner on the reader that has the same map scale as your map. In this case it is 1:24,000 meters. Each line on this grid represents 100 meters.

7. Place the **UTM Corner Grid Reader** on the map sheet with the corner touching the area of interest (e.g., your school, house, park, etc.).

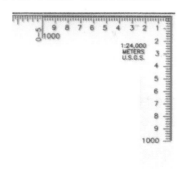

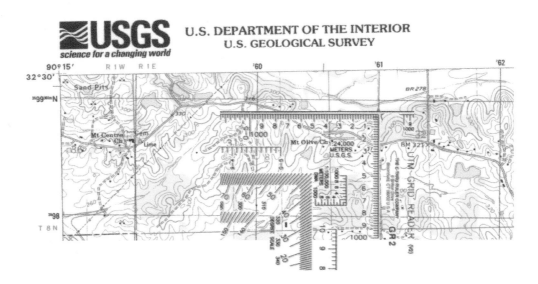

8.   In this case, the easting value (**x**) appears to be located on the vertical easting line, therefore the easting coordinate is 761,000 meters.

9.   For the northing value (**y**), locate the horizontal gridline south of the feature. It appears that this line intersects on the first hash mark below 8 (which stands for 800). This hash mark on the grid ruler indicates 20 meters. With this information, the northing coordinate is 3,598,820 meters.

Grid: Zone 15
761000E 3598820N

10.   Use the **UTM Grid Reader** to identify 6 to 10 landscape features from the map sheet. Use the *S1U3L5 UTM Grid Coordinate Recording Log* to record the **x** and **y** values of each feature.

Lesson 4: UTM Grid Coordinate Recording Log

UTM Grid Coordinate Recording Log

Student Name: _____

UTM Zone: _____

Feature Name	Easting x Value (Step #4)	Northing y Value (Step #5)	Step #8 x	Step #9 y (100ths + 10ths)	UTM x (Easting)	UTM y (Northing)
Example: Campsite	^{7}61 761,000m	^{35}98 3,598,000m	0	820	761,000 E	3,598,820 N

Lesson 4: Map Coordinate Systems Lesson Review

Key Terms
Use the lesson or glossary provided in the back of the book to define each of the following terms.

1. Spatial reference systems

2. Coordinate reference systems

3. Geographic Coordinate System

4. UTM

5. Easting

6. Northing

7. State Plane

Global Concepts
Use the information from the lesson to answer the following questions. Use complete sentences for your answers.

8. List one pro and one con for using the Geographic Coordinate System.

10. How is the UTM map divided up? Where does Zone 1 start?

---

11. What is the benefit behind using UTM zones?

12. Name the units of measurement used in UTM zones.

13. Will you ever have a negative number using UTM? Explain why or why not.

14. The State Plane coordinate system is said to be more reliable than other projections. Why?

15. Name the most common unit of measurement used in State Plane.

Let's Talk About It...
Answer the following question and share the responses with your instructor and classmates.

16. If you were a GIS project manager working for the state of Utah and your project focused in on a small city in southern part of the state, which coordinate system might you use for your project? Be prepared to explain your answer.

17. Why is it difficult for State Plane data to be shared between different zones within a state or across different state boundaries?

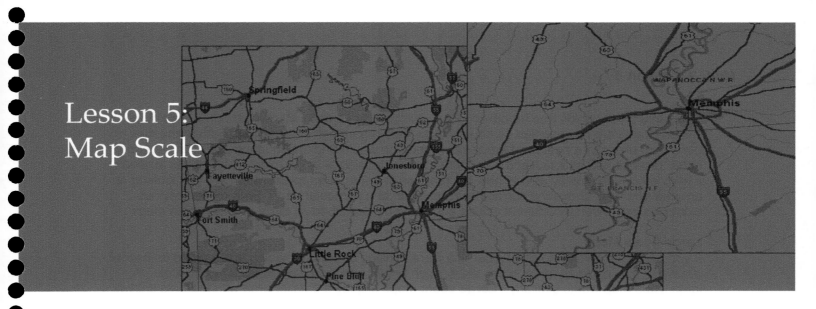

Lesson 5: Map Scale

Previous lessons focused methods to represent the Earth with 3D and 2D models. The final piece of information to complete these models is to explain the difference in size between the Earth and its model. Whether GIS analysts or cartographers are using very large maps or smaller, portable maps, both represent an object many times larger than the map, and this difference must be noted. The relationship between the Earth and a map is called scale.

Model cars (the size that fit in the palm of your hand) are many times smaller than the car that they represent. You may have noticed that some models note their scale, so that audiences know the relationship between the size of the model and the size of the represented car. These are what are called scale models and were made to resemble a full sized product. Similarly, you can have a scaled map of the state you wish to study. When using maps, scales are crucial. If you are looking at a road map, if point A is only an inch from point B, how far is that distance on the ground? 20 miles? 50 miles? or 100 miles? If you are looking at two points on a city map that are only an inch apart, how far apart are they in reality? 20 feet? 2 blocks? or 2 miles?

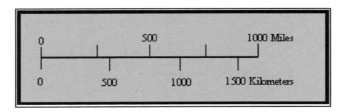

Map scales such as these are found on many maps. What other scale styles have you seen?

A **map scale** is the relationship between distance on the map and the portion of the earth that it represents. This means that if the scale at the bottom of the map reads 1:24000 would mean that any distance on the map is 1/24000th the distance on the ground.

Depending on the map scale, geographic objects on some maps may be represented as points, but on other maps with a different scale, those same objects may be represented as polygons. An example of this would be Atlanta, Georgia. On a world map Atlanta, as well as other major United States cities, may be represented as points. On a Georgia state map however, Atlanta may be represented as a polygon.

The same theory holds true for other geographic objects as well. Many rivers on a map of the United States are represented as equally wide lines regardless of the actual width of the river. On a map of an area near the Mississippi River, you would find that in many places, it could be better represented as a polygon – in some cases with islands in the middle. The scale in which a map is created directly effects how the features in a map are represented. Are you looking at an area from well above the surface or zooming in to a particular part? Both of the previous examples are driven by different scales and could be described as large or small scale maps.

Large Scale Maps
If you had been given photos of a car you were thinking about purchasing and wanted to inspect a scratch, you could zoom in to see the scratch better, which would display a larger scale. A large scale map works similarly. **Large scale** maps display a smaller area with features shown in greater detail.

An example of a large scale map could be one that a realtor shows clients at a scale where individual houses in a certain neighborhood are visible. If the realtor had just given them a city map, most likely all that the clients would have seen were dots representing these locations. With this large scale map, they are able to not only see the addresses but they can also get an idea of how large, as well as the actual shapes of the lots are. They would also be able to see how far away other amenities are such as schools and shopping centers.

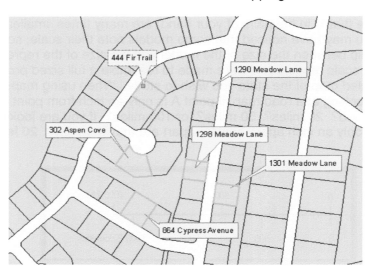

This represents a sample real estate map that would be
supplied to a client. The scale here would be 1 inch = 250 feet.

Small Scale Maps
If your goal this summer was to drive from San Diego, California to Savannah, Georgia, would a large scale map of San Diego be your first choice for a map? Probably not. That large scale map would most like show you the best route to leave the city but after that you would be lost. A small scale map of the United States would probably be your

Introduction to GIS and RS Concepts, version 7

best resource. **Small scale maps** show more area than large scale maps and therefore, would be your best bet for a cross country trip. However, because they show more area, they tend to show less geographic detail of landscape features.

On a small scale map, each individual entity appears smaller due to the large area being mapped. For instance, if you look at North Carolina on a World Map, it is possible that only the cities of Charlotte and Raleigh will be shown. Are they the only cities in North Carolina? Of course not! Since they are among the largest in the state (one is the state capital and the other has a large population), they are the only ones listed at that scale. If you were to use a state map, there may be hundreds of cities and towns including these two listed.

A small scale map means that a bigger geographic area is represented on the map. As a result of the area covered, some features may appear either very small or not appear on the map at all. For instance, many small scale maps only show major roads such as Interstates as opposed to smaller two lane highways.

Location and Representation
Map scales are usually displayed in or near the legend area of a map which is usually found on the periphery of the map. This allows for the user to quickly reference the scope of the map being viewed. There is usually one of three ways that map scale is displayed on a map; as a bar scale, a verbal scale or as a representative fraction.

The **bar scale**, also known as a visual scale, is a line or bar that has tick marks for units of distance. There are many different ways that bar scales can be displayed. These differences are generally aesthetically different, but all are read similarly. The bar scale is especially important because it remains accurate when a map is enlarged or reduced.

A **verbal scale** explains the scale in words: "one inch represents 2,000 feet." To use this scale, you would need to measure the distance in a map and calculate the distance on the ground based on the information given in the scale.

Different Ways that Map Scale Can Be Displayed	
Visual (bar) scale	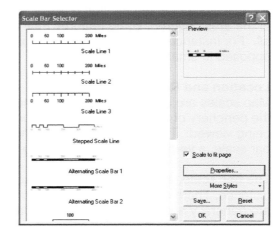
Verbal scale	**One inch represents 500 miles**
Representative fraction	**1/24000 or 1:24000** **Map Scale = Map Distance / Ground Distance** **(in same units)** **= 1 / 24000**

The **representative fraction** is a ratio, such as 1:24,000, in which the numerator (1) represents units on the map and the denominator (24,000) represents units on the ground; in the example of 1:24,000 scale, one unit (any unit -- feet, millimeters, miles, etc.) on the map represents 24,000 of the same units on the ground. The United States Geological Survey (USGS) uses the representative fraction for the topographic maps that they produce.

Map Scales in GIS
In a GIS, map scale is also important when creating map layouts. The maps produced in a GIS must adhere to the same principles as drawn maps and must display their relationship to the Earth. In ArcMap, when a layout is being created, you simply need to insert a scale bar. These can be represented as a scale line or a scale bar and ArcMap has several different styles to choose from. When the map is printed, the user will be able to know exactly how the earth is being represented in the map.

ArcMap also has a feature that allows users to specify a scale for a map. This feature is useful when trying to create a map that has the same scale as another map or when creating multiple maps that should have the same scale. Why is this important? In order to compare "apples to apples" or in this case, states, each must be viewed at the same scale to properly compare the two. Consider the following examples of Texas and Rhode Island.

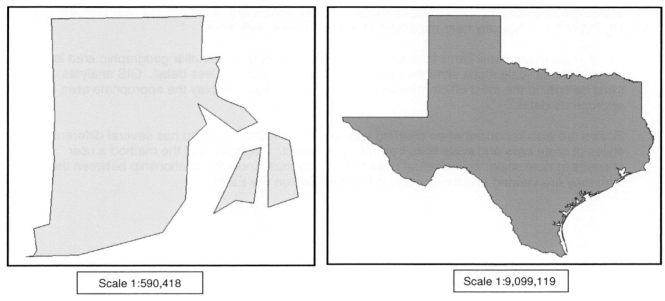

Scale 1:590,418   Scale 1:9,099,119

In ArcMap when "zoomed to layer", both appear to be rather large. Can you tell which state has more area?

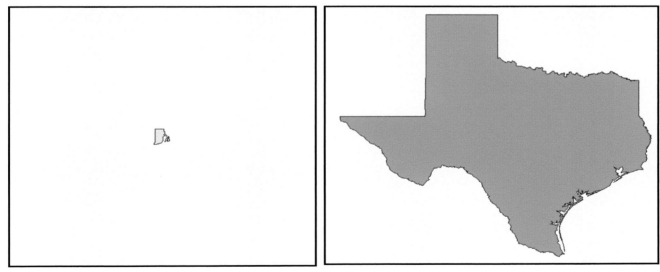

Both maps are now using the same scale of **1:9,099,119** so that Texas may be shown at its full extent. Which scale selection provides the best means of being able to compare the two states?

This feature is also useful for creating maps similar to other organizations. For instance, USGS commonly uses 1:24000 maps for topographic maps; a user could specify the same scale for their own map for comparison.

Conclusion

Map scales are used to express how distance on a map compares to the actual distance on the ground. There are various styles and types of scales that can be used whether they are a bar scale, verbal scale, or representative fraction. These scales are usually found near or in the legend which is typically near the edges of a map for easy reference.

Large scale maps show items in much more detail. As a result, a smaller geographic area is shown. Small scale maps show much more area and show much less detail. GIS analysts must determine the most efficient scale for a map to effectively display the appropriate area and appropriate detail.

Scales are also important when creating map layouts in a GIS. ArcMap has several different styles of scale bars and scale lines from which to select. Regardless of the method a user chooses to note scale, an individual reading the map must know the relationship between the map they are viewing and the corresponding distance on the Earth.

Lesson 5: Map Scale Exercise

In this lesson, you were introduced to the concept of map scale as it pertains to reading and interpreting a map. Use the maps the maps supplied in the appendices of this book to answer the following questions about map scale.

Working with scale
The scale of the map determines the distance on the map and the corresponding distance on the ground. Different maps have different scales, but one map can only have one scale!

1.   Look at the legend for the **shaded relief map**. What is the scale of this map? How many different ways is this map scale defined?

2.   Now look at the **road map**. What is the scale of this map? How many different ways is this map scale defined?

3.   Look at the **topographic map**. What is the scale of this map? How many different ways is this map scale defined?

Using the distance scale
4.   Using your ruler, measure the distance on the **road map** between the State Capitol and the town of Sandy. What is the length on the ruler?

5.   Using the distance scale on the road map, determine how many miles it is from the State Capitol to Sandy.

6.   Now locate the State Capitol on the shaded relief map. Using your ruler, measure the distance on the shaded relief map between the State Capitol and the town of Sandy. What is the length on the ruler?

7.   Using the distance scale on the shaded relief map, determine how many miles it is from the State Capitol to Sandy.

8.   Your answers should be the same! Can you explain why?

Lesson 5: Map Scale Lesson Review

Key Terms
Use the lesson or glossary provided in the back of the book to define each of the following terms.

1. Map Scale

2. Large Scale Map

3. Small Scale Map

4. Bar Scale

5. Verbal Scale

6. Representative Fraction

Global Concepts
Use the information from the lesson to answer the following questions. Use complete sentences for your answers.

7. Explain how is it possible for a feature to be represented as a point on one type of map and then as a polygon on another.

8. Provide one example of an application of a large scale map.

9.   Provide one example of an application of a small scale map.

10.   Draw and label the three different types of map scales.

Let's Talk About It...
Answer the following question and share the responses with your instructor and classmates.

11.   Why are scales so important? Have you ever seen a map that did not have one?

12.   If you were going to a nearby state park, which type of map would you select to get there and which type would you use once you were in the state park?

Lesson 6:
Map Types and Map Essentials

The most effective maps are created with purpose and are easily read by their intended audience. Although maps are applied to a variety of topics and industries, the best maps and GIS analyses begin with similar questions that require thoughtful consideration. When these questions are answered sufficiently, the result is a map that communicates clearly and effectively and is a powerful presentation tool. Some of these questions will address how features in the map appear; some of these questions will focus on font sizes and other graphic elements; and some may focus on map layout. As these questions are answered, the map will become more distinct as a map with specific uses.

Early in the unit, maps were categorized as physical, political, or cultural based on their content, but maps are used for many different purposes. Some of those are specific uses like navigation, while some are more general like viewing locations of countries, which have many applications. Often maps are associated more by their purpose, or their type, than the category in which they fit.

How many different types of maps can you name? Most will be able to name a road map. Some might have even thought of a weather map. Check your own list versus this list of some of the most commonly used types of maps:

- **Climatic Maps**
- **Resource Maps**
- **Shaded Relief Maps**
- **Road Maps**
- **Topographic Maps**
- **Relative Location Maps**

Climatic Maps

Climatic maps typically use a graduated color style that displays weather information such as rainfall, temperature, and humidity. In maps like the one shown on this page, different rainfall levels are placed into categories, and each category has a different color representation on the map.

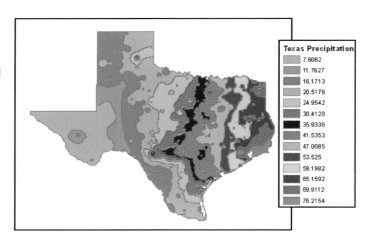

Resource Maps

Resource maps provide locations of natural resources such as oil, gas, uranium, and coal. These may include information that can be found on the surface or below the surface of the Earth. The example shown on this page shows coal beds and mined areas around Pittsburg, Pennsylvania.

These are specialized thematic maps that are used by many entities depending on their content. Examples would be an oil company using a map to decide where to drill or a geologist investigating mineral distribution.

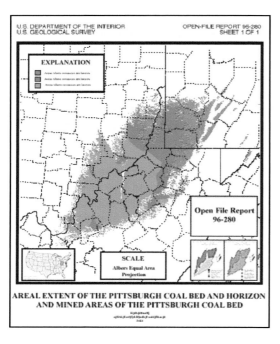

Shaded Relief Maps

Shaded relief maps emphasize change in elevation, or relief, such as the mountains, valleys, rivers, lakes, or the ocean floor, giving the flat map a three dimensional appearance. This look is created showing how an area looks with sunlight shining on it from a particular direction.

Some shaded relief maps, like the one shown on this page from the USGS, also show the locations of cities. Using the legend for this type of map, you can also find out which towns are the largest and which are the smallest.

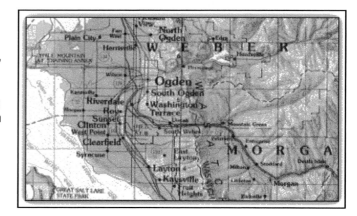

Road Maps
Road maps show people means of travel from one place to another. They also show some physical features, such as mountains and rivers if they are relevant to travel, and political features, such as cities and towns. Road maps, generally, distinguish between major highways and smaller local roads.

Topographic Maps
Topographic maps show measured changes in elevation often represented as contour lines. Topo maps, as they are often called, also can show data such as temperature and elevations. Topographic maps use contour lines to show elevation (vertical changes above sea level) or depth (vertical changes below water) by joining points of equal elevation above (or below) a specified reference. Think of a contour line as an imaginary line that takes any path necessary to maintain constant elevation. People frequently use topographic maps when hiking. Architects and engineers use topographic maps to find the ideal location for buildings and roads. There is a topographic map like this for every part of the United States, including one for where you live.

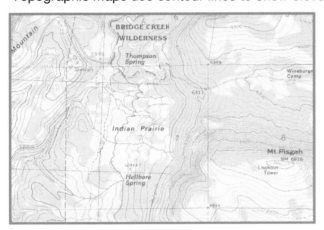

This is a topo map of the Bridge Creek Wilderness area in central Oregon. How could a map like this be helpful if you were hiking in this area?

Relative Location Maps
Relative location maps show a location in relation to a feature's surroundings. These maps are generally not based on a scale to the Earth but rather based on measurements relative to other objects in the map. For example, the map at a local mall is not based on geographic coordinates but is more likely based on the engineering schematics based upon building measurements. These maps are most useful if readers are in the mapped area or familiar with it. When navigating using a relative location map, readers rely on directions such as "next to", "beside", or "two units down" rather than precise coordinates.

Map Essentials
If you were given a piece of paper and a pen, could you sketch a map of your current location? Chances are you can. The questions that should pop into your mind when you look at a map is, "when was it created?", "who created it?", "which direction is north?" To keep the reader from having to ask these or similar questions, there are at least seven items that should be present in every map that is created. These are:

- **Title**
- **Symbols**
- **Legend**
- **Author**
- **North Arrow/Compass Rose**

- **Scale**

- **Date**

The **title** of a map is a concise description of the map or its purpose. If a title is not included on a map, it may be impossible for the reader to figure out what type of data is represented on the map. In instances, where maps are included in presentations, map titles are crucial in capturing a reader's attention. Subtitles may also be included to further enhance the reader's understanding of the map.

Symbols on a map illustrate and distinguish the **points, lines,** and **polygons** which represent features on the Earth. First, regarding symbol selection, it is important to choose a symbol that best represents the characteristics of the feature being emphasized. Notice in the example, the vegetation in the building received a green plant and the water fountains received a blue circle. Second, as you learned in the previous lesson, the scale of a map effects how features are represented on it. On a small scale map of the United States, a city would most likely be listed as a point. Another example at a different scale might be the locations of water fountains on a map of a campus.

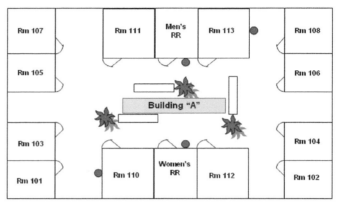

A sample campus map showing locations of water fountains as points in blue. Can you name any other items that could be listed as points that are not currently shown on this map?

Just because the point is described as point, does not mean that only a small circle can signify it. There are numerous symbols that can be used for points. ArcMap comes with a database of literally hundreds of point symbols to select from. A sampling of these point symbols is shown below:

| School 1 | Airport | Hospital 1 | Toilets | Disabled 1 | No Smoking |

How many of these have you seen before? With ArcMap, it is easy to incorporate point symbols such as these in your map layouts. Not only do you have the capability to change their size (just as you can change font sizes), you can usually change the color of the symbols as well.

Lines are another way that features are represented on a map. The function of the line may determine the thickness of the line's symbol. An example of a feature that could be represented as a line would be a road on a map. In road maps, the different types of roads may be distinguished by the thickness of the line symbol. Just as is the case with point features, ArcMap has hundreds of line symbols to select from. As with point symbols in ArcMap, you have the ability to select the color of the lines that you will use in your map layouts, which is also useful to distinguish a line feature. A sampling of these line symbols is shown below:

| Highway | Railroad | Stream or Creek | Cold Front 2 | Residential Street | City Limits Outline |

A third type of feature symbolized in a map is a polygon. Polygons can take the form of any enclosed shape. An example of a feature type that could be represented as a polygon would be state boundaries on a map of the United States or the boundaries of a city zoo on a local map. In ArcMap with polygons, you will have many choices to select from that vary in fill color, outline color, or patterns. A sampling of these is shown below:

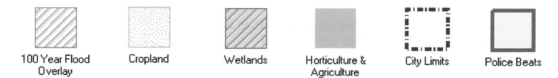

| 100 Year Flood Overlay | Cropland | Wetlands | Horticulture & Agriculture | City Limits | Police Beats |

With these polygons, you can alter the colors used just as you can with the points and line symbols. One other benefit of using polygons in ArcMap is that you can also edit the borders of the polygons making them thicker, thinner, or you can even have a hollow border around your polygons.

With points, lines, and polygons, it is important to keep your final map in mind when editing features. You may have to change the size of the point features, or the borders on the polygons in order to keep your map readable. Colors are also another important decision to consider when editing the features. In the end the goal should be a map that clearly illustrates the map's central idea(s) and is also appealing to the reader.

The **legend** provides a list of all the features – points, lines, and polygons - used on a map and provide a description of what each feature represents. Without a legend, a map would be difficult, if not impossible, to understand.

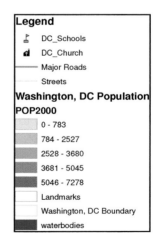

Another map essential is a directional element. Many maps, but not all, are drawn with the top of the page or map oriented north. In order to be completely certain as to which direction the actual map is oriented, either a **north arrow** or a **compass rose** should be included. What is the difference? A north arrow is a simple arrow that only designates which direction is north on a map – although the appearance of the north arrow may vary. Many applications created in ArcMap will have arrows like the one shown on the right to designate north.

A **compass rose**, however, shows more than just north. A compass rose can show both cardinal (N, S, W, and E), ordinal (NE, NW, SW, and SE) directions on a map, or in some cases may show all 360° of direction.

Map scale shows how the size of the map corresponds to the size of mapped features in real life. Map scale can be represented as a visual scale, a verbal scale, or as a representative fraction.

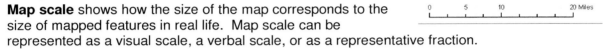

Two additional map essentials are the **author** and the **date**. It is important that readers have some reference to the map's author because the map may reach extended audiences, without the author's direct input. Knowing the date that the map was created is also important. Unless you need historical information, you will usually want to access to the most up-to-date information as possible.

Can a map be produced without all of these essentials? Yes, but, without the seven essentials listed here, the validity of that map would be in question to the person using it.

Conclusion

Maps are made for many different reasons and purposes. There are many different types of maps including topographic, road, physical, economic, and political maps. From maps that show you where something is located, or how far it is from point "A" to point "B", to others made to show elevation or terrain in an area, it may take a little research to find the map best fit for your situation.

Maps use point, line and polygon symbols to represent geographic features on the Earth. In a program like ArcMap, the symbols used can be edited in color and, in some cases; more realistic symbols can be added to the map to make it even more comprehendible. To be most effective, maps need to contain a title, appropriate symbols, a legend, an author, the date, a scale and north arrow.

Lesson 6: Map Types and Map Essentials Assessment Exercise

Whatever the form, symbols are used to effectively represent geographic features on a map. In this lesson activity, you will gain experience reading the symbols of various types of maps.

Topographic Maps:
Topographic (or topographical) maps are representations of a portion of the Earth that illustrate the surface using contour lines. These contour lines show the shape of the Earth's surface. Figure 1 is an example of a surface of the Earth showing the shape of the landscape from a perspective view. Figure 2 is a map of the same area displaying the elevation contours of this landscape.

Figure 1: Perspective view of Earth surface   Figure 2: Map view of surface with elevation contour lines

(Source: *http://erg.usgs.gov/isb/pubs/booklets/symbols/reading.html*)

Notice that the steeper the elevation, the closer the elevation contours appear on the map. Conversely, more widely spaced contour lines reflect areas with lower grade slopes. Each contour line represents equal elevation so no two contour lines ever cross. To help read the contour data, wider contour lines called *index contours* are inserted every fourth or fifth contour. Topographic maps in relatively flat areas usually have contour intervals of 10 feet or less because the extent of elevation from highest to lowest point is relatively small. Conversely, maps illustrating mountainous areas may have contour intervals of 100 feet or more because the extent of elevation is greater from highest to lowest point. Bathymetric contours are shown to represent the surface of the ocean bottom.

In addition to contour lines, topographic maps also include symbols that represent geographic features. These symbols may vary slightly from map to map, but all are chosen based on how they resemble the real features on the ground that they represent. In order to effectively read a "topo" map, you must be able to interpret these symbols.

Topo maps use color to effectively represent features such as representing vegetation in green, water in blue, and more developed areas in gray or red. Map features are represented using *points, lines* or *polygons*. Some buildings may be represented using small black squares while larger buildings may be represented using their actual shapes. Meanwhile, other densely located features may not be represented individually, but may be represented with an area tint. Different line styles and colors are used to represent features with brown lines reserved for the elevation contours. Bathymetric contours are represented in blue or black.

Elevation Symbols
The following symbols are typically used to represent elevation data on topographic maps produced by the United States Geological Survey (USGS). Notice that some of the features include handwritten symbols for older topo maps.

CONTROL DATA AND MONUMENTS

Aerial photograph roll and frame number*	3 - 20

Horizontal control

Third order or better, permanent mark	Neace △ Neace ⊕
With third order or better elevation	BM △ 45.1 ⊕ P.K. BM 45.1
Checked spot elevation	△ 19.5
Coincident with section corner	△ Cactus ⊕ Cactus
Unmonumented*	+

Vertical control

Third order or better, with tablet	BM × 16.3
Third order or better, recoverable mark	× 120.0
Bench mark at found section corner	BM + 18.6
Spot elevation	× 5.3

Boundary monument

With tablet	BM □ 21.6 BM ⊕ 71
Without tablet	□ 171.3
With number and elevation	67 □ 301.1
U.S. mineral or location monument	▲

CONTOURS

Topographic

Intermediate	
Index	
Supplementary	
Depression	
Cut; fill	

Bathymetric

Intermediate	
Index	
Primary	
Index Primary	
Supplementary	

Boundary Symbols

The following symbols are typically used to represent boundary data on topographic maps produced by the USGS:

BOUNDARIES

National	
State or territorial	
County or equivalent	
Civil township or equivalent	
Incorporated city or equivalent	
Park, reservation, or monument	
Small park	

LAND SURVEY SYSTEMS

U.S. Public Land Survey System

Township or range line	
Location doubtful	
Section line	
Location doubtful	
Found section corner; found closing corner	
Witness corner; meander corner	WC MC

Other land surveys

Township or range line	
Section line	
Land grant or mining claim; monument	
Fence line	

Symbols for Land Surface Features

The following symbols are typically used to represent land surface features on topographic maps produced by the USGS:

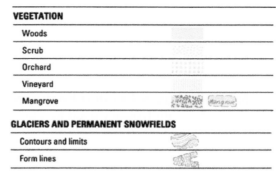

SURFACE FEATURES	
Levee	
Sand or mud area, dunes, or shifting sand	
Intricate surface area	
Gravel beach or glacial moraine	
Tailings pond	

MINES AND CAVES	
Quarry or open pit mine	
Gravel, sand, clay, or borrow pit	
Mine tunnel or cave entrance	
Prospect; mine shaft	
Mine dump	
Tailings	

VEGETATION	
Woods	
Scrub	
Orchard	
Vineyard	
Mangrove	

GLACIERS AND PERMANENT SNOWFIELDS	
Contours and limits	
Form lines	

Symbols for Water Features

The following symbols are typically used to represent water features on topographic maps produced by the USGS:

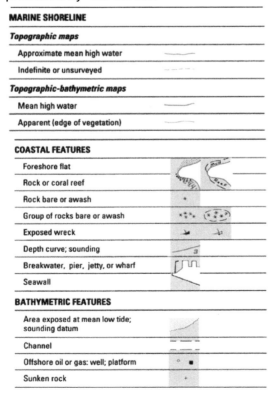

MARINE SHORELINE	
Topographic maps	
Approximate mean high water	
Indefinite or unsurveyed	
Topographic-bathymetric maps	
Mean high water	
Apparent (edge of vegetation)	

COASTAL FEATURES	
Foreshore flat	
Rock or coral reef	
Rock bare or awash	
Group of rocks bare or awash	
Exposed wreck	
Depth curve; sounding	
Breakwater, pier, jetty, or wharf	
Seawall	

BATHYMETRIC FEATURES	
Area exposed at mean low tide; sounding datum	
Channel	
Offshore oil or gas; well; platform	
Sunken rock	

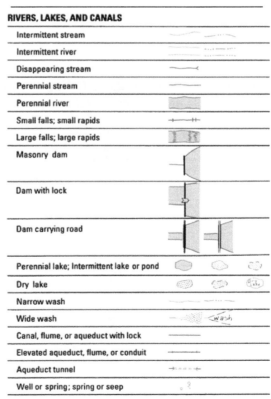

RIVERS, LAKES, AND CANALS	
Intermittent stream	
Intermittent river	
Disappearing stream	
Perennial stream	
Perennial river	
Small falls; small rapids	
Large falls; large rapids	
Masonry dam	
Dam with lock	
Dam carrying road	
Perennial lake; Intermittent lake or pond	
Dry lake	
Narrow wash	
Wide wash	
Canal, flume, or aqueduct with lock	
Elevated aqueduct, flume, or conduit	
Aqueduct tunnel	
Well or spring; spring or seep	

SUBMERGED AREAS AND BOGS

Marsh or swamp	
Submerged marsh or swamp	
Wooded marsh or swamp	
Submerged wooded marsh or swamp	
Rice field	*Rice*
Land subject to inundation	

Symbols for Buildings & Related Features

The following symbols are typically used to represent buildings and other related features on topographic maps produced by the USGS:

BUILDINGS AND RELATED FEATURES

Building	
School; church	
Built-up Area	
Racetrack	
Airport	
Landing strip	
Well (other than water); windmill	
Tanks	
Covered reservoir	
Gaging station	
Landmark object (feature as labeled)	
Campground; picnic area	
Cemetery: small; large	Cem

Symbols for Roads, Railroads & Other Features

The following symbols are typically used to represent roads, railroads and other features on topographic maps produced by the USGS:

ROADS AND RELATED FEATURES

Roads on Provisional edition maps are not classified as primary, secondary, or light duty. They are all symbolized as light duty roads.

Primary highway	
Secondary highway	
Light duty road	
Unimproved road	
Trail	
Dual highway	
Dual highway with median strip	
Road under construction	*U. C.*
Underpass; overpass	
Bridge	
Drawbridge	
Tunnel	

RAILROADS AND RELATED FEATURES

Standard gauge single track; station	
Standard gauge multiple track	
Abandoned	
Under construction	
Narrow gauge single track	
Narrow gauge multiple track	
Railroad in street	
Juxtaposition	
Roundhouse and turntable	

TRANSMISSION LINES AND PIPELINES

Power transmission line: pole; tower	
Telephone line	*Telephone*
Aboveground oil or gas pipeline	
Underground oil or gas pipeline	*Pipeline*

Road Map:

Roads maps are used to illustrate how to travel from one place to another. In addition to roads, road maps also typically show physical features and political features. Road maps can depict larger geographic areas with less feature detail; or smaller geographic areas with greater feature detail. For example, road maps that depict whole states may not include small, rural streets, while a county map may include them. Symbols used for a state road map may resemble the following:

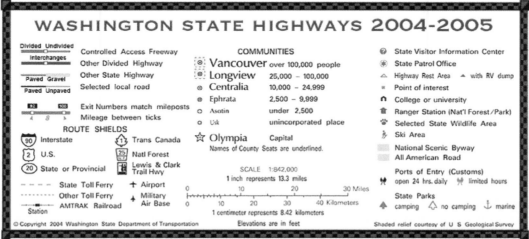

(Source: Washington State Department of Transportation, State Highway Map,
http://www.wsdot.wa.gov/Communications/Map/view.htm)

Symbols used for a county road map may resemble the following:

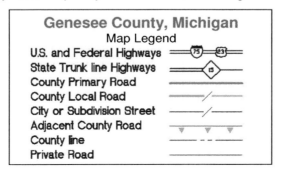

(Source: Genesee County Road Commission, Genesee County, MI,
http://www.gcrc.org/MapLegend.html)

Shaded Relief Map:

You can use a shaded relief map to find locations of towns and cities, but the primary use of this type of map is to illustrate the shape of the land by showing how the area looks if sunlight is shining from a certain direction. Figure 3 is a shaded relief map of a section of the Grand Canyon.

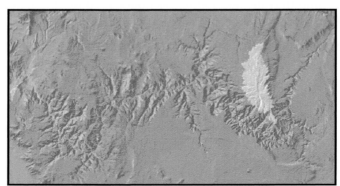

Figure 3: Shaded relief map of a section of the Grand Canyon
(Source: *http://pubs.usgs.gov/of/1999/of99-011/grand_canyon_m.gif*)

Lesson Activity:

(The activities in this unit are adapted with permission from the USGS What Do Maps Show?, http://erg.usgs.gov/isb/pubs/teachers-packets/mapshow/index.html)

Topographic Maps

Use the the illustrations in this section to answer the following questions.

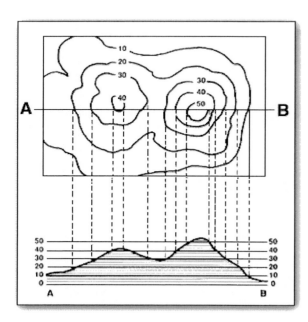

The top of this drawing is a contour map of the hills illustrated along the bottom. On this map, the vertical distance between each contour line is 10 feet.

1. Which is higher, hill A or hill B?

2. Which is steeper, hill A or hill B?

3. How many feet of elevation are there between contour lines?

4. How high is hill A?

   How high is hill B?

5. Are the contour lines closer together on hill A or hill B?

Look at the picture below. It shows a river valley and several nearby hills.

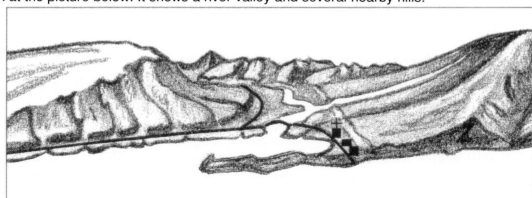

6.   On the illustration, locate the following things:

   - A church
   - A bridge over the river
   - An oceanside cliff
   - A stream that flows into the main river
   - A hill that rises steeply on one side and more smoothly on the other.

Here is a topographic map of the same place. Find the items you located on the illustration o the topographic map.

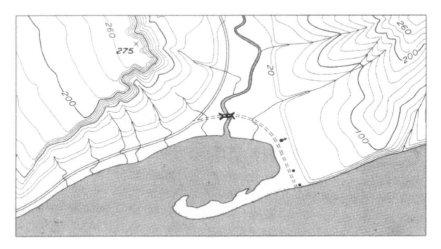

7.   Circle the symbol for a church.   Draw a church symbol here.

8.   Put a square around the map symbol for a bridge. Draw a bridge symbol here.

9. Put an **X** on the oceanside cliff. What is the elevation of the contour line at the top of that cliff?

10. Locate a stream that flows into the main river. Draw a pencil line down that stream. Put an **X** where the stream joins the main river.

Tell how you might use a topographic map if you were selecting:
11. A route for a hike.

12. The best location for an airport.

13. A route for a new road.

Use the **Topographic Map Poster** in the appendices (or a STARS Series 1 Materials Kit if available) to answer the following questions:

1. Which is steeper, the area to the north or south of the police firing range located in the upper right corner of the map?

2. Find and draw a symbol for a school.

3. How high is Ensign Peak?

4. What is the elevation of the jeep trail west of Ensign Peak?

5. Draw the symbol for railroad.

6. What is the color used for rivers or creeks?

7. What is the contour interval of this topographic map?

8. What is the approximate elevation of the State Capitol?

9. Would you be walking uphill or downhill to go from the State capitol to Pioneer Park?

10. Suppose you lived by Fremont School. Find at least three ways you could get from your house to the State capitol.

11. List some of the features you would see along the way on one of the routes.

Road Maps

Road maps show people how to travel from one place to another. Use the **Road Map Poster** in the appendices (or a STARS Series 1 Materials Kit if available) to answer the following questions:

1. Find and draw the map symbol for an Interstate highway.

2. Find and draw the map symbol for a State highway.

3. Locate an Interstate highway.   What is its name?

4. Locate a State highway. What is its name?

5. Why do you think there are so few roads north and east of the city?

6. What colors are highways?

7. The areas around a city are shown in color on the map. What color is used for Salt Lake City?

8. What color is used for bodies of water?

Shaded Relief Maps
Shaded relief maps are designed to show the shape of the land. Use the **Shaded Relief Map Poster** in the appendices (or a STARS Series 1 Materials Kit if available) to answer the following questions:

1. Locate a canyon on the map. What is its name?

2. What features do you see at the bottom of a canyon?

3. Why do you think these features are at the bottom of the canyon?

4. Find and draw the symbol for airport.

5. Which direction is the airport from the State Capitol?

6. The legend shows city size. Using the legend, the population of Salt Lake City is between _____ and _____.

7. The population of Ogden is between _____ and _____.

8. What is the name of a town with a population of 500 to 1,000?

9. What are the major colors on the map and what does each color represent?

Working with Direction
You can use north, south, east, and west to locate where one place on a map is in relationship to another.

Use the **Shaded Relief Map Poster** in the appendices (or a STARS Series 1 Materials Kit if available) to answer the following questions:

1. On this map, which town is farthest north?

2. Which town is farthest south?

3. Which towns are farthest east?

4. Which town is farthest west?

5. If you were traveling from Ogden to Sandy, what direction would you travel?

6. If you were traveling from Farmington Bay to Willard Reservoir, what direction would you travel?

7. Echo Reservoir is _____ of Salt Lake City.

8. Great Salt Lake State Park is _____ of Salt Lake City.

9. Deer Creek Lake State Recreation Area is _____ of Salt Lake City.

10. West Valley City is _____ of Salt Lake City.

11. What direction do the Wasatch Mountains run?

12. What direction does Echo Canyon run?

Map Critique
Study the map below and answer the questions that follow it.

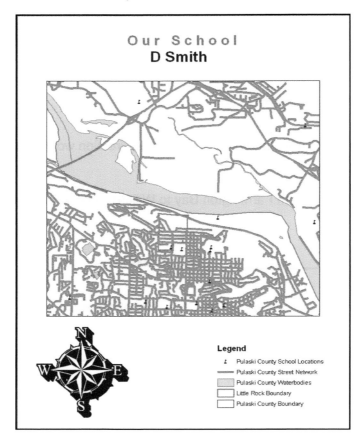

1. There are at least two map essentials missing from this map. Name them.

2. How does not having them effect your interpretation of the map?

3. List at least three things about this map that you would change. Explain each one.

Lesson 6: Map Types and Map Essentials Lesson Review

Key Terms
Use the lesson or glossary provided in the back of the book to define each of the following terms.

1. Symbols

2. Features

Match that Map
Match the map with the description given. Write the letter of the best answer in the blank.

Map Type	Description
_____ 3. Climatic	a. great for planning a trip
_____ 4. Resource	b. good for displaying weather related information
_____ 5. Shaded Relief	c. usually found in malls
_____ 6. Road	d. hikers often use this type of map
_____ 7. Topographic	e. these maps almost appear 3D
_____ 8. Relative location	f. diamond mines would be documented on this type of map

Global Concepts
Use the information from the lesson to answer the following questions. Use complete sentences for your answers.

9. List the three ways that symbols are displayed on a map. List one example of each.

10.   Explain how using a symbol instead of a black dot for a point would be beneficial.

11.   List the seven essential items that all maps should have.

12.   Explain how a north arrow differs from a compass rose.

Let's Talk About It...
Answer the following question and share the responses with your instructor and classmates.

13.   You are writing a research paper on the growth of the city of Los Angeles, California. You only found one map thus far and although it looks recent, it does not have a date or an author listed on it. Would you use it? Why or why not?

Map Critique

Scenario: You own a limousine company that has decided that its clients would benefit from maps showing proximity of area hotels to the main airport. The clients with children would also like to know where local parks and local hospitals are located. The marketing department came up this map. Study it and answer the questions that follow.

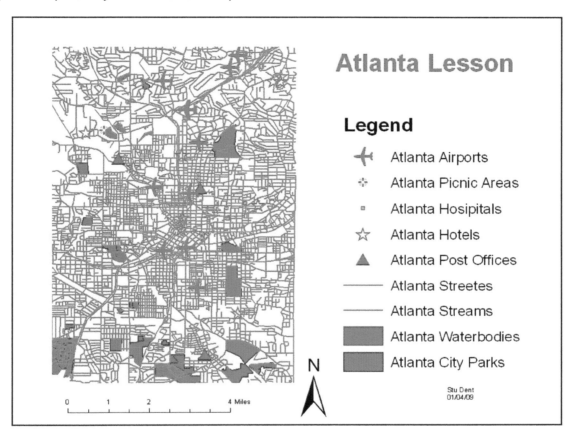

1. The map covers the exact area that your company drives but is the map large enough to answer their questions? Why or why not?

2. Produce a list of at least five things that you would have done differently if you had created this map. Provide explanations for each.

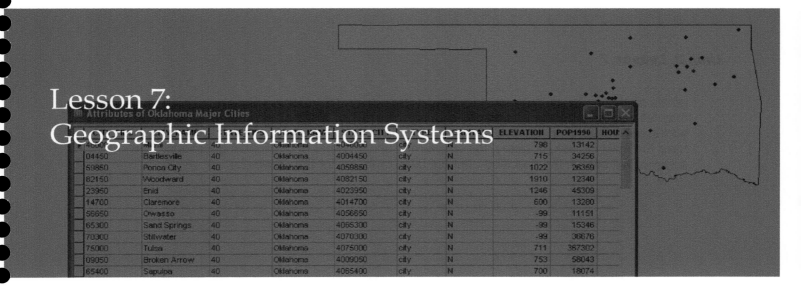

The lessons up to this point have dealt with representing the Earth on either a flat surface or a globe. Any time you look at a map or a globe, you have to draw your own conclusions as to why something is where it is located, how far two places are from one another or what the best route is from one state to the next. Wouldn't it be nice to be able to plug a question in to the computer and have it give you an answer? Like how far is it from Orlando, Florida to New York City? Or where are the major cities with populations over 500,000 people located? Questions like these can be answered easily with a GIS. So what is GIS? Why is it used?

Geographic Information Systems

Geographic Information Systems (GIS) is a combination of data, computer software and hardware system for storing, sorting, processing, and displaying geographic information. Just as with any software system, the information given to a user by a GIS is only as good as the information put into it.

In its most basic usage, a GIS is used to clearly visualize your theme of interest. What does this mean? When you look at a printed map, you must view all of the data that is included on that map from Interstates, to rest areas, to state parks, to farm and ranch roads, business routes, and more. In a GIS, you can display only the data that you are interested in viewing. If, for example, in a GIS, you have a map that displays all major roads in your area, and you are only interested in studying the interstates, then you can turn "off" all layers that are not interstates. This is just one of the many perks of using a GIS.

The use of Geographic Information Systems has changed the cartographic process in one major way; it has added an analytical element to the traditional cartographic process of data collection, data manipulation, map production, and map reproduction.

Traditional maps contain a set amount of data. How is a GIS different?

Behind every feature that is added to a GIS, there is a table of information about that feature. Each piece of descriptive information in these tables is known as an **attribute**, and all of the attributes collectively make up an attribute table. An attribute table for rivers might tell you the names of the rivers being displayed, the length of the river, the depth of the river, and their

points of origin. An attribute table for road networks from the Census Bureau's TIGER line data may tell you describe the names of the streets, the length of the street segment, the range of addresses included on that line segment, and the direction of each segment. With a GIS you have the ability to ask, or query, the software to find answers to geospatial questions. The software will search the specified attribute table or tables for the answer(s).

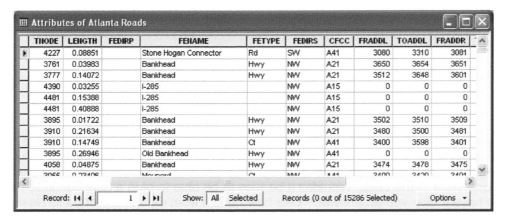

TNODE	LENGTH	FEDIRP	FENAME	FETYPE	FEDIRS	CFCC	FRADDL	TOADDL	FRADDR
4227	0.08851		Stone Hogan Connector	Rd	SW	A41	3080	3310	3081
3761	0.03983		Bankhead	Hwy	NW	A21	3650	3654	3651
3777	0.14072		Bankhead	Hwy	NW	A21	3512	3648	3601
4390	0.03255		I-285		NW	A15	0	0	0
4481	0.15388		I-285		NW	A15	0	0	0
4481	0.40888		I-285		NW	A15	0	0	0
3895	0.01722		Bankhead	Hwy	NW	A21	3502	3510	3509
3910	0.21634		Bankhead	Hwy	NW	A21	3480	3500	3481
3910	0.14749		Bankhead	Ct	NW	A41	3400	3598	3401
3895	0.26946		Old Bankhead	Hwy	NW	A41	0	0	0
4058	0.04875		Bankhead	Hwy	NW	A21	3474	3478	3475
3055	0.23408		Maynord	Ct	NW	A41	3400	3420	3401

This attribute table from the Census Bureau contains various types of information such as feature name, feature type, direction, and address ranges for roads in Atlanta, GA.

Why Do We Use GIS?

There are many reasons that we use GIS. First is the ability to overlay different information to find interesting geographic relationships. These two maps alone look interesting, but can they tell us more about volcanoes and their relationship to the Earth? When the information is layered (as seen on the following page), it makes a much more informative layout. Knowing the locations of the current eruptions that have taken place versus the locations of the tectonic plates may even make it possible to better predict where future earthquakes may occur.

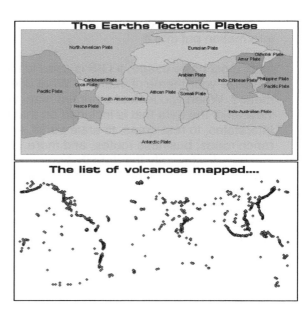

Does there appear to be a relationship between the Earth's tectonic plates and the locations of volcanoes?

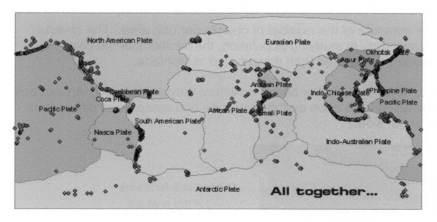

Why are volcanoes located in certain locations around the world? This graphic shows that many of them lie on the edges of the Earth's tectonic plates.

Discovering Geographic Relationships

Another reason to use GIS is to learn not only determine where landscape features are, but also helps us to determine why they are there. Knowing how geographic features are interrelated is a very powerful tool. Understanding spatial relationships unlocks a tremendous amount of information that can help us make knowledgeable decisions regarding the Earth and its resources. Why are so many major cities located near water? How many houses are located within a mile of a factory? You can use GIS to not only visualize why things are where they are, but analyze them also.

We can describe these spatial relationships in many ways including:

- **Proximity** – refers to the quality of being near something. It is measured by a distance between two features or objects. Consider the proximity between hotels and restaurants. A motorist may ask, "Why are certain types of restaurants located near hotels?" Using that same relationship, a pedestrian may ask, "Why are restaurants located far from hotels?" The measured distance between restaurants and hotels did not change, the application and the perception of the user did.

- **Spacing** – refers to objects' distances from one another. Spacing may be uniform or random. Spacing may be studied to determine where a string of burglaries has taken place in order to predict where the next burglary might occur. Spacing does not have to apply to the entire feature group. For instance, when taking into account the points within the city on the map below, the points appear to be uniform, however if the map displayed all of the points within the county, they might not seem to be uniform.

Legend
- • Burglaries
- — Streets
- City Boundary

Spacing can be used to determine not only where incidences like burglaries have occurred, but can also be used to detect if patterns exist that might likely help predict where the next burglary might occur.

- **Density** – refers to the measure of the number of objects per unit area. The objects may be clustered or dispersed. In studying population, the locations of the cities in Colorado may be considered clustered than those found in Montana.

- **Orientation** – refers to how objects differ based on where they are located. An example of orientation would be when vegetation on the north side of a steep slope differs from the vegetation on the south side.

The four cellular towers
(▲) in this area are
located at various altitudes
(white areas are the highest
elevations). Those that are
oriented at the highest
elevations may provide the
best reception. What
reason do you see for this?

- **Diffusion** – refers to how objects move or are dispersed as time progresses. Numbers or statistics can tell someone how much or how far something has spread such as a disease or even something like fire ants. Being able to show them on a map brings another important dimension to the numbers.

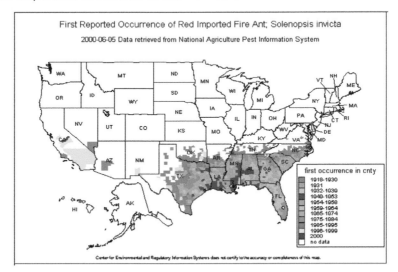

Many species of fire ants have entered our nation through southern ports.
How far have they spread?

- **Association** – refers to the spatial relationship that exists among elements that occur at the same location. When one spatial pattern may be partially or totally related to another spatial pattern, they are said to have a strong association. Historically, the locations of cemeteries have been strongly associated with streams. Why is this true? The ground near the streams is usually softer than ground away from streams.

This map shows an association between older cemeteries and streams. Why do you think so many cemeteries were built so close to streams?

(Historically the ground was softer near streams.)

GIS Data Types

The third reason GIS is used is the fact that the Earth's geography changes faster than maps can. Often, by the time a street map for a major city has been printed, new streets have already been added that were not there by print time. GIS enables you to view the most up-to-date information or edit data for manual updates. GIS relies on two formats of data to make this possible: **vector** and **raster**.

Vector data

No matter how the relationships or any other features are displayed on a map in a GIS, digital data, just like traditional maps, also uses points, lines, and polygons. When data is displayed in a shapefile format as a point, line, or polygon, this can be described as vector data. Individual points or vertices, in the case of lines and polygons, correspond with a specific location (for instance latitude and longitude coordinates) on the surface of the Earth.

Vector data is displayed as points, lines, and polygons such as those represented here.

Why is this important and how is vector data applied to GIS? To understand vector data more clearly, let's look at a shapefile from its creation. What can be said about the shape to the right? It could be described as the state of Nevada or as the "Silver State" due to its vast number of silver mines, or as the seventh largest state (in area), or the 36th state in the union, or many other descriptions. The shape can also represent numeric data like total population, statewide average income, number of state parks, etc. The one thing that all of these descriptors have in common is that they apply to an area with a distinct border. Data that represents phenomena with distinct boundaries is **discrete data**.[1]

Polygons such as this one of the state of Nevada contain discrete data such as statewide average income or a total population.

Vector data allows GIS users to create a point, line, or polygon and apply a specific value to that shape. The polygon to the right can be all of the descriptions above, and the descriptions apply to all area within that boundary. For this reason, the data is considered homogenous.

When creating this data, a GIS user will input the vertices of the polygon (in sequential order) to define its shape. In the image to the right, the state boundary of Nevada is shown with the vertices set by the data's creator. So in this example, the creator may have begun by setting the first vertex in the northwest corner of the state and continued placing vertices in order until the shape of Nevada was complete.

A GIS will store the information about the shape of this polygon based on the location of each individual vertex. The table on the right displays 16 of the vertices used to create the Nevada polygon. If any of the points was considered to be incorrect, it would be easy to edit its location using the table shown or by simply dragging the vertex to its correct location.

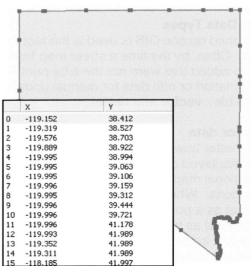

	X	Y
0	-119.152	38.412
1	-119.319	38.527
2	-119.576	38.703
3	-119.889	38.922
4	-119.995	38.994
5	-119.995	39.063
6	-119.995	39.106
7	-119.996	39.159
8	-119.995	39.312
9	-119.996	39.444
10	-119.996	39.721
11	-119.996	41.178
12	-119.993	41.989
13	-119.352	41.989
14	-119.311	41.989
15	-118.185	41.997

---

[1]Tasha Wade and Shelly Sommer, *A to Z GIS* (ESRI Press, 2006) 58.

Just as the example above uses a polygon, vector based shapefiles use the same process to create points and lines. The point and line features will have the same advantages as the polygon. They will represent locations that correspond with a specific set of coordinates and are easily edited. Editing shapefiles is a common GIS need. Light poles are added to city utility infrastructure represented as point files, roads are added to street networks represented as lines, and borders are adjusted in political districts represented as polygons. Using vector based shapefiles to represent points, lines, and polygons give GIS analysts an accurate and editable format to represent features.

Raster data

Another way to display data in a GIS is through the use of raster data. Recall the previous example of the state boundary of Nevada. Various, homogenous data, where all areas shared the same value, are best displayed with vector data. All points within the Nevada border share the same name, governor, statewide education rank, etc. However, not all data is homogenous; some data is continuous and varies at any point across a surface. Elevation data, temperature, and rainfall amounts have a unique value at any point on the surface of the Earth, and that variation is extremely important in analyzing our world. Raster data is a format that represents this type of continuous data. Continuous data does not have distinct boundaries like the data in our Nevada example; therefore, it is not efficient to have a polygon representing all of these varied values.

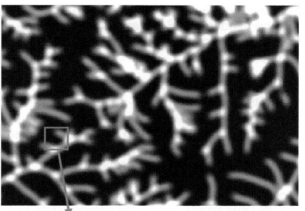

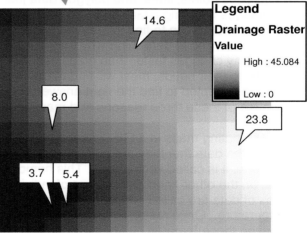

Raster data is made of a grid of equally sized cells that have unique numeric values. In a GIS each cell will represent a specified area on the ground. For example, each cell may represent an area that is 30 meters by 30 meters, or if more accuracy is required, a smaller cell size may be used. The GIS displays these numeric values as colors. A lower numeric value may display a light color while a higher value may display a darker color. In the example to the right, elevation is

Top image: typical raster data file
Bottom image: When zoomed in, the raster data cells are visible. As shown in the callout boxes, each cell has its own unique value.

represented using raster data. The darker numbers indicate a higher altitude and conversely, the lighter numbers represent lower altitudes. Each cell has its own unique value.

There are many file formats used in GIS that display raster data. Common types are grid data and various image formats like tiff and jpeg. Elevations and temperature are data commonly represented as grids. This data is often created from sample data or remotely sensed data which is covered in later units. Imagery, also remotely sensed, is similarly represented in raster format.

Raster data can be quickly acquired through satellites and other remote technologies, or also can be quickly created from a set of sample data. For instance, if temperature readings at various locations are available, then a GIS can estimate the values over an entire surface. So as the Earth changes, gathered or created raster data formats can be used in a GIS to easily show changes across a surface over short or long time periods.

Imagery, like this satellite photo of Washington, DC, is another example of raster data.

Support of raster and vector data formats in a GIS gives analysts the means to represent any type of data collected or observed on the Earth. The data is not only easily displayed but in most cases is also quickly gathered. The ability to create dynamic maps quickly and efficiently is essential to good decision making in spatial problems.

Conclusion
A GIS uses points, lines, and polygon data just as traditional maps do. There are many advantages for using a GIS as opposed to traditional maps such as; the ability to overlay different information to find interesting geographic relationships, to discover where landscape features are and to determine why they are there, and for the simple fact that the Earth's geography changes faster than maps can. GIS allows users to build the maps and applications specific to their needs, most will find many more advantages unique to their situation. Because they are expected to find the right answer(s) quickly, project managers use GIS as a tool because of these three reasons and many more.

When creating maps in a GIS, users can find data sources that are very current or, if necessary, update the data that they are currently using. When data is displayed as points, lines, and polygons, it is called vector data. This homogenous type of data is fairly easy to edit. Raster data, on the other hand, is made up of individual cells, each with their own unique value. These types of data are called continuous data and can include data such as imagery and temperature. Once raster or vector data features are added and displayed, relationships between these features can be determined using relational descriptors like proximity, spacing, density, orientation, diffusion, and association.

Lesson 7: GIS Fundamentals & Geographic Relationships

Directions: Review the following maps and answer the corresponding questions.

Map 1:

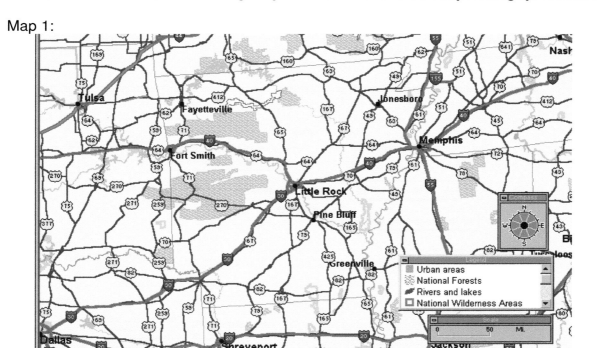

Source: Rand McNally Tripmaker

1.   Name one map entity that is represented as points on Map 1.

2.   Name two map entities that are represented as lines on Map 1.

3.   Name two map entities that are represented as polygons on Map 1.

4.   What can be said about the **association** between cities and water bodies according to Map 1?

5. Describe the **proximity** of Pine Bluff to Little Rock using Map 1.

6. Describe the **spacing** of national forests around the cities of Memphis and Little Rock.

Map 2:

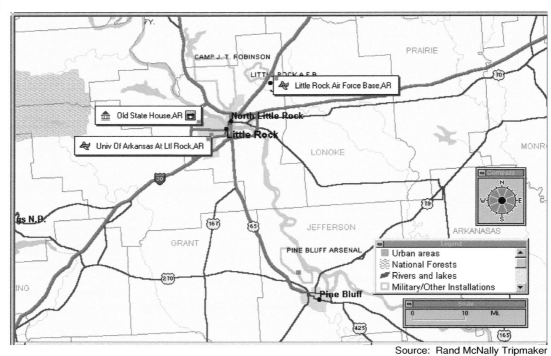

Source: Rand McNally Tripmaker

Use Map 2 to answer the following questions.

7. List at least one point, one line and one polygon representations that have changed from Map 1 to Map 2.

8. Evaluate the **association** of route of state hwy 65 and major rivers.

9. Describe how the **density** of tourist attractions relates to urban and rural areas. (answer will need to have "clustered or dispersed" as opposed to close

10. Describe the **proximity** of Pine Bluff to Little Rock using Map 2.

Use Lesson 7 manual or your PowerPoint notes to answer the following.

11. Identify the three major reasons given in the lesson as to why GIS is used.

12. Every feature layer of points, lines, and polygons that is added to a GIS has an attribute table with descriptive information attached to it. Create a list of at least five attributes that could be contained in your city's attribute table.

Raster vs. Vector

(a) Read the following scenarios and apply your knowledge of raster versus vectors to determine if the answer is raster or vector.

(b) Apply this knowledge to describe the possible appearance for that layer (for example: a polygon, a point file, a continuous grid, etc.).

13. I have a layer file with all of the fire hydrants in a city.

14. I need a map layer that will show temperatures in Montana.

15. I need a map layer that will display all national parks within the southwest.

16. I have a map layer that displays rainfall amounts in Hawaii.

17. I need to create a map layer to show the locations of all cities in the US that have theme parks.

18. I have a map layer that displays all rivers in the state of Louisiana.

19. I have a map layer that displays the average state temperature for Arizona for the month of December.

Lesson 7: Geographic Information Systems Lesson Review

Key Terms

Use the lesson or glossary provided in the back of the book to define each of the following terms.

1.   Geographic Information Systems

2.   Attributes

3.   Vector data

4.   Raster data

5.   Discrete data

6.   Continuous data

Matching

Match the geographic relationship with the description given. Write the letter of the best answer in the blank.

Map Type	Description
_____ 6. Association	a. may be uniform or random – how features are distributed
_____ 7. Density	b. relationship that exists between elements that occur at the same location
_____ 8. Diffusion	c. measured by approximate distance between features
_____ 9. Orientation	d. how clustered or dispersed features are in an area
_____ 10. Proximity	e. how objects differ based on where they are located
_____ 11. Spacing	f. how objects are dispersed over time

Global Concepts

Use the information from the lesson to answer the following questions. Use complete sentences for your answers.

12.   When using a GIS, you have the ability to turn off layers that are not needed. How is this different than traditional maps and why would this be beneficial for the GIS user?

13.   Provide two examples of information that can be found in an attribute table.

14.   List three reasons why GIS is used.

15.   Vector data is described as homogenous because if you have a polygon, the entire polygon has the same value regardless if you are selecting an area on the border of it or in the center. How is this different than data that is a raster data?

Let's Talk About It...
Answer the following question and share the responses with your instructor and classmates.

16.   Why is data such as rainfall represented as raster data instead of vector data?

17.   This lesson covers three advantages to using a GIS over traditional mapping however, there are many others. List at least one additional advantage and explain how it is an advantage.

Introduction to Remote Sensing Concepts

Introduction to Geographic Information Systems and Remote Sensing *Concepts*

Unit Four

Lesson 1:
Remote Sensing & Aerial Photography

Have you ever noticed the viewpoint that is used to create maps? When you look at a map, you are looking at the area mapped as if you were above it looking down. For hundreds of years maps have been drawn of many places – showing the area from the orthogonal (overhead) viewpoint, but it wasn't until the last century that we actually got to see what the Earth looked like from above its surface. We now have the capability to attach cameras and other specialized equipment to objects that are many miles above the earth to gather imagery and data about the Earth. GIS allows us to use that remotely sensed imagery and data for creating very accurate maps.

What is Remote Sensing?
Remote Sensing is the observation of objects, from a distance. In geospatial studies, the term remote sensing refers to gathering information about our world from a distance. With respect to the acquisition of Earth imagery, remote sensing captures these observations using a photographic or scanning device. The two most common methods of capturing images by remote sensing are by aerial photography and satellite imagery.

Anytime that you look at something from a distance, you are involved in remote sensing. However, to gather data about the Earth and its surface, we use these two data collection strategies to do so. By using information collected from these sources, areas of the world can be studied without having to travel to that area.

One type of collecting remotely sensed data is about taking "pictures" of the Earth. There are, however, other techniques of remotely capturing information that go beyond traditional methods. Remote sensing can be split into two categories based on how the information is collected: passive remote sensing and active remote sensing.

Passive Remote Sensing
Passive Remote Sensing uses sensors or cameras that produce imagery gathered via energy or light that is naturally reflected from objects on the ground. This is similar to taking a picture with a camera (without a flash). These sensors or cameras used in passive remote sensing are carried by airplanes or satellites. They can capture imagery of the world as it appears to the human eye and can also collect imagery that reveals data the human eye is incapable of seeing.

Panchromatic Photography
The most common types of photos are taken with panchromatic film. **Panchromatic images** are grayscale photos of land areas taken in the visible portion of the electromagnetic spectrum. Images are usually black and white or grayscale and show features as we would see them if we were flying overhead.

Aerial photographs allow you to see land features with good resolution. It is fairly easy to recognize features once you have a good point of reference. One major drawback to aerial photos is the limited area that can be obtained in a single picture.

Aerial photographs are produced by exposing film to solar energy reflected from Earth. The intensity or brightness of a feature will correspondingly be displayed as bright in the image.

Infrared Photography
Infrared photography is used to capture infrared portions of the electromagnetic spectrum. Human eyes see certain types of energy which we use to view the world around us. These images will appear as a false color image. The human eye is incapable of seeing this infrared portion. A false color image will replace infrared areas with a color the human eye is capable of seeing. In the photograph to the right, the infrared portions of the image appear red. The darker colors of red represent more intense infrared energy.

The image on the slide is a color-infrared of Niagara Falls. Notice that the trees are red in the image. This is characteristic of healthy vegetation in a color-infrared image.

Color-infrared imagery is useful for distinguishing between healthy and diseased vegetation, delineating bodies of water, and penetrating atmospheric haze. Color-infrared film, which records energy from portions of the electromagnetic spectrum invisible to the human eye, was developed to detect camouflaged military objects in the 1940's. In a color-infrared (also known as false-color) photograph, near-infrared light reflected from the scene appears as red, red appears as green, green as blue, and blue as black. Color-infrared film is useful for distinguishing between healthy and diseased vegetation, for delineating bodies of water, and for penetrating atmospheric haze and cloud cover.

Infrared images show objects' levels of emitted infrared radiation. Black-and-white and color-infrared films are used today in both high- and low-altitude aerial photography. Why not use regular color film? Natural-color film is used less often because it is affected by atmospheric haze and cloud cover.

The type of infrared photography that is used depends entirely upon what type of study that is being conducted.

Airborne Scanners

Cameras of various types are used to take aerial photographs. Although cameras have also been carried on spacecraft such as the Space Shuttle, satellites more frequently use electronic scanners to record ground scenes in digital form.

Sources of Aerial Photography

Aerial-photography firms are typically small, local operations. The industry generates about $800 million a year in revenues, and clients are largely regional governments and engineering firms that are mapping cities and boundaries for further development and public works.

This image shows the AVIRIS scanner and how it captures imagery as it passes over the Earth.

There are many private commercial aerial survey companies in the United States. Most work with both traditional cameras and digital scanners to produce aerial photography products.

The **National High Altitude Photography (NHAP)** program was a federal image acquisition program that ran from 1980 to 1989. NHAP photographs were taken simultaneously with two cameras, one containing black-and-white film, the other, color infrared. The NHAP aircraft flew at 40,000 feet. These photographs cover more area, but show less detail, than NAPP photographs. An NHAP black-and-white photograph covers about 129 square miles, and because of the longer focal length in the second camera, an NHAP color-infrared photograph covers about 68 square miles.

This image is a color infrared NHAP image of the Golden Gate Bridge.

The **National Aerial Photography Program (NAPP)** was established in 1987 to coordinate aerial photography for the United States among federal and state agencies. NAPP photographs are used for such purposes as mapping, resource planning, engineering, land use planning, and agricultural monitoring. Private citizens also purchase them for hunting, hiking, and other recreational uses. Taken from aircraft flying nominally at 20,000 feet above the terrain, each NAPP photograph covers about 32 square miles.

Another popular US imagery program is the **National Agriculture Imagery Program (NAIP)**. The primary purpose of the NAIP is to acquire peak growing season "leaf on" imagery, and deliver this imagery to United States Department of Agriculture (USDA) County Service Centers. NAIP imagery is acquired at a one meter ground sample distance (GSD) with a horizontal accuracy that matches within five meters of a reference ortho image. Earlier imagery gathered has an accuracy of two meter GSD.

The **United State Geological Survey (USGS)** has used aerial photographs for many decades, although the availability of photographs from the earliest years is limited. Most recent photographs are from programs covering all 50 States and were taken with predominantly black-and-white film, with some instances of color-infrared use.

USGS Digital Ortho Quadrangle (DOQ) are images that have a uniform scale. Traditional aerial imagery has distortions due to the angle of the camera and the tilt of the Earth. You cannot use a traditional aerial image to measure distances due to this distortion. DOQ's, on the other hand, have been "orthorectified" using ground elevation data to correct displacements caused by differences in terrain relief and camera tilt. The process gives an orthophoto the accuracy of a map. Because imagery shows the texture of the ground in much

greater detail than do maps, digital orthophotos are useful for updating maps and for studying surface features not necessarily visible in maps. The USGS produces digital orthophotos for map revision and for computer analysis using geographic information systems. Coverage of each photo aligns with the "USGS 7.5 minute" topographic map sequence of the entire United States.

The **Earth Resources Observation Systems (EROS) Data Center** of the USGS has an extensive archive of USGS photos and also negatives of aerial photographs from other federal agencies, including the Bureau of Land Management, the Bureau of Reclamation, the Bureau of Indian Affairs, NASA, and the Armed Forces. EROS is an important resource for historical imagery.

NASA uses aerial photographs for research and to test remote sensing techniques and instruments. These photographs, available in various formats, are taken from altitudes of a few thousand feet up to more than 60,000 feet. NASA aerial photographs may be available in black and white, natural color, or color infrared.

This image was taken of Hurricane Elena by the Space Shuttle over the Gulf of Mexico in September, 1985.

Active Remote Sensing
In contrast to passive remote sensing, active remote sensing uses electronically generated signals or light waves that bounce off targets. Depending on the type of device used, the sensors use the time and shape of the signal to determine what the area looks like. One type of active remote sensing is **Doppler radar**. Doppler radar is used for various applications, the most popular being for weather.

Side-looking airborne radar (SLAR) instruments on aircraft or satellites generate their own energy, which is recorded on being reflected back to them from the ground. This eliminates problems associated with cloud cover and haze. The oblique angle of the "side-looking" instrument yields images that are especially useful in analyzing landforms.

Conclusion

Remote sensing involves studying something from a distance. Two common types of remotely sensed data are aerial photography and satellite imagery. There are two categories of data collection methods for remote sensing: passive remote sensing and active remote sensing. Passive involves collecting information about the Earth just as you would with a camera without a flash. An example of passive remote sensing involves aerial photographs. Aerial photographs are taken using panchromatic (grayscale) and color infrared film; or are captured as digital images. Another method of passive remote sensing is using airborne scanners aboard aircraft to collect aerial photos. Active remote sensing involves using a type of signal that is sent to the desired area and then once that data bounces off of the surface, the way it returns will be recorded and interpreted. The type of remote sensing used will vary based on the desired end result of the project.

Lesson 1: Remote Sensing & Aerial Photography Identifying Features from Aerial Photos Exercise

There are many different sources for aerial photography. Some are only available for purchase from a private aerial survey company that has been commissioned to fly over specific areas for a client's particular needs. There are, however, other sources of aerial photographs that are available online for low or no cost. Fees for this data vary depending on the source, the quality (resolution), and whether the data already exists or must be collected. For a GIS project manager, finding the correct imagery and ordering it are important steps in the project management process.

For this exercise, you will explore an online source for imagery and collect data to document your source for metadata. When you begin working with imagery for use in GIS projects, you will use data that has been purchased and processed specifically for your school.

Using the USGS EarthExplorer Website:

1.   ***Click*** on the ***USGS EarthExplorer*** link on the ***SPACESTARS*** website or navigate to the website – **http://earthexplorer.usgs.gov/**

2.   Under ***1. Enter Search Criteria,*** enter your address in the address text box in the format *Address, City, State.*

3.   Click **Show**.

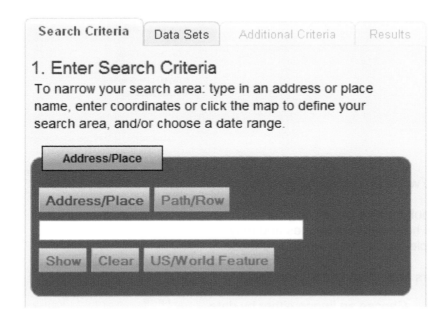

A marker will be placed on the map inset.

4. Select the **Data Sets** Tab

5. Click the ⊞ by **Aerial Photography** to expand the list

6. ***Select* DOQ** by checking the box beside it.

    Although there are other options, a DOQ is the image of preference for this exercise.

7. Click the **Results »** button at the bottom of the Data Sets tab.

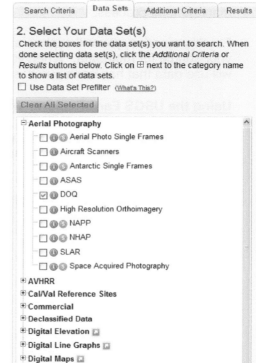

8. The **Results** tab will open displaying your results.

    NOTE: Your results will be based on your search criteria. You will have different images and may even have multiple images from which to choose.

9. A toolbar appears under each image description ⬜🖫📁⚓⊘, Choose an image either by date or appearance by and ***click*** the **Show Metadata and Browse** button.

A window will open displaying a larger image and information about the image.

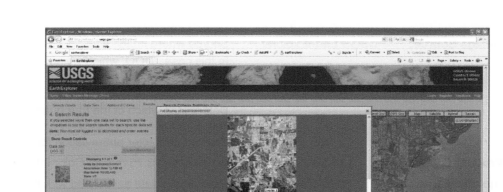

10.   In the display window, ***click*** the [Open New Window] button to open a larger display like the one below:

11.   Scroll down to see the metadata for the image.

There are several fields of information contained in the Results window that need to be recorded in order to create a metadata file after the image has been added to a mapping software document. This will not only verify that this image is from a credible source, it will give others important information should someone in the future want to download this DOQ themselves.

Image Location:	Ridgeland, MS	Area in the image
Image Source:	USGS EarthExplorer	Source where image was obtained from
Image Type:	DOQ	In this case: DOQ, which we set in our search criteria
Map Name:	Ridgeland	DOQ quadrangle map name
Quadrant:	SE	Determines which of the four quadrangles is in this image
State:	MS	The two character abbreviation for the state
Product Group:	3.75 Min CIR	In this case: 3.75 minute Color Infrared image
Acquisition Date:	1996/02/12	Most recent data associated with this DOQ
Resolution:	1 (meter)	Measure of sharpness of the image
Entity ID:	DI00000000891997	Used to uniquely ID data

12.   **Record** the information from your **local DOQ** in the table included on the **Identifying Features Lesson Activity** sheet on the following page.

At this point, if you were a project manager, and your AOI was included in this DOQ, you would download the image to your computer. Because these files are very large, for this class, you will not have to do that at this time. Instead, as part of this assignment, follow the directions below to capture this image to be printed and turned in with your Lesson Activity worksheet.

13.   Scroll to the top of the screen so that you can see the image preview..

14.   To capture this full image that is larger than the screen, ***right click*** on the image and ***select*** **Save Picture As…** from the menu provided.

15.   **Save** the picture to your **student folder** as **Local Image**.

16.   ***Open*** a Word document.

17.   From the **Main Menu**, ***select*** **Insert**, then **Picture**, **From File**.

18.   ***Maneuver*** to your **student folder** and ***select*** **Local Image**.

19.   ***Resize*** the image (if necessary) so that it completely fits on one page with room above it to key information.

20.   ***Document*** the **Image Location**, **Image Source**, and **Entity ID** by keying them above the image.

21.   ***Save*** the Microsoft Word document as **Local_Image_XX.doc** (where **XX** is your initials) in your **student folder.**

22.   ***Print*** one copy of the **Local Image** and ***staple*** it to the back of the **S1U4L1 Activity Worksheet** to turn in to your instructor.

Introduction to GIS and RS Concepts, version 7

Lesson 1: Aerial Photos Activity Aerial Photo Worksheet

A common responsibility for a GIS project manager is to locate the correct imagery for a task. Certain details such as the cost of the imagery and the date of the imagery should be taken into consideration before the project manager decides on an image source. Once this has been determined, the project manager will not only obtain the imagery, but also collect important data, as shown below, so that a metadata file can be created once the image is brought in to a mapping software.

Directions: Follow the steps outlined in the lesson instructions to find an aerial photograph of your local area using the website, USGS EarthExplorer. Once you have located a DOQ of your area, fill in the following table with the requested information. Attach the Local DOQ printed with this exercise and turn it in to your instructor.

Student Name: _____

Image Location:	
Image Source:	
Image Type:	
Map Name:	
Quadrant:	
State:	
Product Group:	
Acquisition Date:	
Resolution:	
Entity ID:	

Lesson 1: Remote Sensing and Aerial Photography Lesson Review

Key Terms
Use the lesson or glossary provided in the back of the book to define each of the following terms.

1. Remote Sensing

2. Passive Remote Sensing

3. Active Remote Sensing

Global Concepts
Use the information from the lesson to answer the following questions. Use complete sentences for your answers.

4. Explain how it can be beneficial to study something from a distance rather than going to that location.

5. Explain how passive remote sensing differs from active remote sensing.

6. Compare and contrast Panchromatic photography versus Infrared photography by listing one pro and one con for each.

7. What type of remote sensing is Doppler radar and what is its most popular application?

8. How are SLAR instruments able to avoid problems with cloud cover and haze?

Sources of Aerial Imagery

Place the letter of the source next to the correct description. Each will be used at least twice.

Description	Source
_____9. NASA	a. National High Altitude Photography
_____10. EROS	
_____11. NHAP	b. National Aerial Photography Program
_____12. NAIP	
_____13. NAPP	c. National Agriculture Imagery Program
_____14. USGS	
_____15. An important resource for historical imagery.	d. United States Geological Survey
_____16. acquires peak growing "leaf on" imagery	
_____17. used two cameras one with black and white film, the other with color infrared film	e. Earth Resources Observation Systems
_____18. capable of collecting imagery from a few thousand to more than 60,000 feet	
_____19. has used aerial photography for many decades	f. National Aeronautics and Space Admin
_____20. created to coordinate aerial photography among federal and state agencies	

Let's Talk About It...

Answer the following question and share the responses with your instructor and classmates.

21. If your job was to access the amount of coastline that had been eroded after a hurricane, which type of remote sensing imagery do you think would be the best one to use and why?

Lesson 2:
Building Data Layers from Aerial Photographs

Have you ever been on an airplane and been able to get a window seat? If you have, chances are, once you were in the air, you were able to recognize some, if not all of the items on the ground. You may discover something about your surroundings that you did not know before - like how many homes in a certain neighborhood have pools or where a golf course is in relation to your home. From streets to buildings to trees, everything can look a little different from the overhead perspective. Looking at a remotely sensed image is similar to being on a flight. It gives you a "snapshot" of a certain area. There are people whose full time job is to interpret remotely sensed imagery. Why do we have a need for this? How is it useful to GIS?

Remote Sensing and GIS

GIS allows users to create maps specific to their needs. Although there are large amounts of data available for the public for free or available for purchase, there still may be times when the specific data needed for a project must be created. If the project is small, the person needing the data may be able to go and collect the data themselves without it taking up too much time or incurring too many expenses. But what if the project required the study of a very large geographic area, or an area that was deemed hazardous due to a recent catastrophe, or may be an

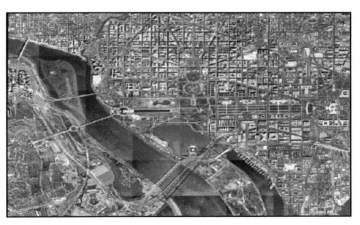

Aerial images, like this one of Washington, D.C., provide a detailed view of the landscape so that geographic features can be easily identified.

area that was too difficult to travel to? It is in times like these that remotely sensed imagery can be useful. **Aerial images** provide a detailed view of the landscape so that geographic features can be easily identified. It is possible to take a remotely sensed image of an area, create various vector data layers by marking all pertinent landmarks – with green dots or points for trees, blue lines for streams, and aqua polygons for ponds for example - and ultimately stacking all of these layers to create a map of that area. This is a very simplistic explanation of a GIS procedure called **heads-up digitizing**. The difference is that in a GIS you are using an image on a screen to trace or mark vector data from – such as points, lines, and polygons. You could record all objects that you see in the image including trees, waterbodies, houses, streets, parks,

parking lots, fire hydrants, and much more. But what if you only wanted to focus on trees and waterbodies? Or you wanted to focus only on houses and streets?

Data Layers

Any time you draw stars for a certain feature, like for example, schools, you are creating a **data layer**. If all you trace are houses, then you have created a house data layer. With a GIS you can have many, many layers in a map. Does that defeat the purpose of using a GIS over a printed map? No, one of the many benefits of using a GIS over a paper map is that although you may have many layers in your GIS map, you do not have to display them all at the same time. And even if you initially used a remotely sensed image to trace these features and create these layers, displaying the image in the GIS you have created is completely optional.

 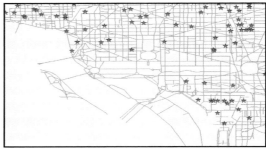

Both maps show the locations of schools in the Washington, D.C. area. The map on the left shows schools and the tif of Washington, D.C.; the one on the right is showing only the schools and the streets in the D.C. area. With a GIS you have the ability to be able to view the information that is most important to you.

With a GIS, you have the ability to view the information that is most important to you. Extra layers can be added, deleted, or simply turned off depending on their relevance to your particular study.

Legend

▯	DC Hospitals
▯	DC Churches
★	DC Schools
—	DC Streets

Map of Washington, D.C. showing schools, churches, and hospitals. Each of the items listed in the Legend is a data layer.

When you combine data layers to form maps, you have the concept of what a GIS can do.

Conclusion

Remotely sensed data, such as air photos, allows you to view many geographic features. These features can be "copied" from air photos, either manually or by using computers. Each of these features that represent some map entity such as trees, roads, or buildings is called a data layer. Data layers can be combined as needed to form specialized maps. Using this same methodology on a computer is the basis behind GIS.

Lesson 2: Building Data Layers from Aerial Photography Exercise

In this lesson, you are to create a map from an aerial photograph by identifying features in the photos and marking them on a transparency sheet.

Directions:

1.	***Print*** out one of the aerial photographs of your community that you used in the previous lesson.

2.	***Obtain*** at least **four** transparency sheets, and **four** different colored transparency markers.

3.	***Select*** **one** marker and ***choose*** **one** type of geographic feature, or data layer, that you want to map.

	Examples include: roads, houses, businesses, trees, recreational facilities, schools, waterbodies, etc.

4.	***Lay*** the transparency on top of the aerial photograph. You may want to secure the transparency with a piece of tape to keep it from slipping.

5.	Using the colored transparency marker, ***draw*** all of the features that you see of your data layer. For example, trace all of the roads or lakes that you see in the photo with the marker.

6.	When finished, ***remove*** this transparency, or data layer, and set it aside.

7.	***Place*** another clean transparency sheet on the same position on top of the aerial photograph.

8.	***Select*** a different color marker and a different feature, draw all of the features that you see for this data layer.

9.	***Repeat*** these steps two more times using a new transparency and new marker each time.

Building Maps using Aerial Photography:

Use a clean piece of 8.5 x 11 inch printer paper as the backdrop; ***overlay*** the transparencies that you have created one layer at a time. How is this similar to a GIS?

Note: You may want to stack these on an overhead projector if available. Do not use the printer paper if you do this. Simply stack the data layers and look at your "map".

Lesson 2: Building Data Layers from Aerial Photographs
Lesson Review

Key Terms
Use the lesson or glossary provided in the back of the book to define each of the following terms.

1. Aerial images

2. Heads-up digitizing

3. Data layer

Global Concepts
Use the information from the lesson to answer the following questions. Use complete sentences for your answers.

4. Explain when might it be considered advantageous to use remotely sensed images?

5. How can aerial images help a GIS user create data files? What is this GIS procedure called?

Let's Talk About It...
Answer the following question and share the responses with your instructor and classmates.

6. What can you do with data layers in a GIS that you cannot do with the data you see when using traditional maps?

Lesson 3:
Remote Sensing & Satellite Imagery

Remote sensing involves studying something from a distance. In the last lesson, aerial photography was discussed as a source of remote sensing, but it is not the only source for imagery. Did you know that satellites have been recording information about the Earth since 1960? What are some of the satellites called? How do we get that data? What can we do with the data once we get it?

Satellite Sensors

Although in the first satellites were run and maintained be some form of government, now there are also commercial businesses that collect and distribute imagery.

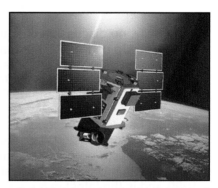

QuickBird is owned by a commercial business (DigitalGlobe) and the imagery collected from Quickbird is highly accurate and contain high resolution. How accurate are QuickBird's images? The resolution in the imagery that QuickBird delivers is accurate to less than one meter. This allows you to clearly focus in on an area of the ground of less than two square feet. That is not too bad for a satellite that is approximately 450km (280 miles) above the Earth! QuickBird is not the only satellite or imaging device up there, below are a few others.

QuickBird, launched by Digital Globe in 2001 and is capable of collecting imagery that is accurate within one meter.

Another imaging satellite is the **IKONOS**. IKONOS is a high-resolution satellite that is operated by Space Imaging LLC. Applications of the imagery from IKONOS include natural resource mapping, natural disaster analysis, agricultural/forestry analysis, change detection analysis and other applications.

The first **Systeme Pour l'observation de la Terre** (**SPOT**) satellite, developed by the French Centre National d'Etudes Spatiales(CNES), was launched in early 1986. Four other SPOTs have been launched since then. The SPOT satellite can observe the same area on the globe once every 26 days. It can be programmed from the ground control station and is quite useful for collecting data in a region not directly in the path of the scanner or in the event of a natural or man-made disaster, where timeliness of data acquisition is crucial. It is also very useful in collecting stereo data from which elevation data can be extracted.

Another imaging instrument that is attached to a NASA satellite is **ASTER** (Advanced Spaceborne Thermal Emission and Reflection Radiometer). ASTER is being used to collect detailed information of land surface temperature, reflectance, and elevation.

Spatial Technology And Remote Sensing

SPACESTARS

STARS
Remote Sensing Education

Laboratory for GIS/Remote Sensing Education

The primary observing systems in the tropics are the **Geostationary Operational Environmental Satellites** (**GOES**). These satellites, orbiting the earth at an altitude of about 22,000 miles above the equator, normally provide imagery every 30 minutes, both day and night. With these images, forecasters can estimate the location, size, movement, and intensity of a storm, and analyze its surrounding environment. Because they stay above a fixed spot on the earth's surface, these satellites can continuously provide data on a particular event.

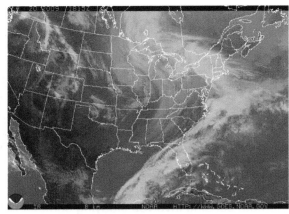

Infrared GOES image of the central and eastern portion of the US. It is easy to distinguish the frontal boundary that is moving off of the east coast.

In the early 80s, India began development of a local **IRS (Indian Remote Sensing Satellite)** program to support their national economy. With the IRS, they felt it would cover many areas such as agriculture, water resources, forestry and ecology, geology, water sheds, marine fisheries, and coastal management.

And finally, no discussion of satellites would be complete without a discussion of the **Landsat** satellites. Up until 1972, the majority of the satellites that were orbiting the Earth were for weather and defense, Landsats, however, were sent to help observe the Earth's terrain. This program was a collaborative effort between National Aeronautics and Space Administration (NASA) and the United States Geological Society (USGS).

The first Landsat, launched in 1972, was launched for experimental global coverage of the Earth's land masses. Landsats 2 through 5 were launched in 1975, 1978, 1982, and 1984. Landsat 5 is still operational many years after its expected life span.
Landsat 7 is a U.S. satellite system that gathers environmental data. It was launched in April, 1999 and is the latest in a series of Landsat satellites launched. Much of the satellite imagery data used for GIS comes from the Landsat 5 and 7 satellites.

Depending on the scope and scale of the project, each of these image sources produces a specific product that satisfies a particular need.

Acquiring Satellite Imagery
Satellites have been acquiring data of the Earth's surface for over 40 years. Today there are literally thousands of satellites above us at any given time. The tasks of these satellites can range from defense, to communications, to exploration. Some are stationary and some are on the move. Because the focus of this lesson is on remote sensing, the satellites and receivers that collect imagery data are the ones that will be our focus. The data, in the form of images that they gather, are not like those taken with conventional film cameras – they more closely represent those taken by digital cameras which store information digitally. Satellites gather data through the use of scanners that collect data from various channels or bands in the electromagnetic spectrum. Remember that image data such as satellite imagery is raster data.

How does the data get to the satellite? The sun emits rays of light, or energy, on our planet everyday. Whenever that energy is sent to the Earth from the sun, some of that energy is reflected while some is absorbed. Different objects on the Earth absorb and reflect energy in different ways. Satellites above our Earth, equipped with sensors attached to them, record energy as it is reflected back from the Earth's surface in a digital format. These numbers are transmitted back to the Earth to ground stations where the data is converted by computers into black and white scale images that look similar to photographs.

Electromagnetic Spectrum

The human eye is a sensor that detects certain ranges of electromagnetic radiation or energy on the Earth. Remote sensing devices serve similar purposes. They detect electromagnetic radiation through a sensor like an eye, and a computer interprets those readings similar to a brain. Each object on the landscape reflects and absorbs light and energy differently. Collecting data in each of these bands helps us identify objects on the ground according to how they handle light and energy. Certain objects show up differently when looking at them at different bands.

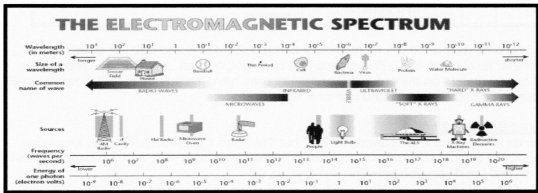

www.lbl.gov/MicroWorlds/ALSTool/EM

The **electromagnetic spectrum** is a scale that displays the arrangement of electromagnetic radiation in terms of energy, wavelength, or frequency. If you look closely at the scale on this page, you will notice that there is only a very small section that is deemed "visible". The energy, or electromagnetic radiation, in this range is what we can see, or detect, with our eyes. Other energy also included on the scale are radio, TV, microwave, infrared, visible, and ultraviolet light. Remote sensing focuses primarily on acquiring image data from the visible, infrared, and microwave radar portions of the electromagnetic spectrum.

Landsat collects data along the electromagnetic spectrum using an Enhanced Thematic Mapper. The Enhanced Thematic Mapper measures reflected light and energy in the visible and invisible parts of the electromagnetic spectrum in seven bands. The data are transmitted to Earth, where they are processed by computers and stored. Each of these seven bands collects information from a certain portion of the electromagnetic spectrum which it is sensitive to and has its own strength for that specific application. For instance, Band 1 is the best of the seven for viewing water depth and penetration.

This chart shows the seven bands and their capabilities:

Band #	Location		Band Focus
Band 1	Visible	**Blue**	Shows water depth
Band 2		**Green**	Shows water turbidity and plant life
Band 3		**Red**	Shows snow cover and distinction between soil and plants
Band 4	Invisible	Near Infrared	Shows clear distinction between land and water
Band 5		Mid Infrared	Shows the moisture content of plants
Band 6		Thermal	Records heat emitted from objects
Band 7		Mid-Infrared	Detects rock structures and mineral deposits

Landsat 7 records data at 30 (for multispectral bands 1,2,3,4,5,7) and 120 (for band 6 or thermal) meter pixel resolution. This means that each dot, or pixel, on the image represents either 30 or 120 meters on the ground. Landsat images have the advantage of showing small scale patterns of landforms and land use over large areas.

TM Infrared Bands 1,2 and 3
These bands show manmade features clearly, but TM Band 1 is particularly good for seeing water depth; TM Band 2 shows water turbidity and different plants well; and TM Band 3 shows snow cover and the distinction between soil and plants.

TM Infrared Bands 4,5 and 7
TM Band 4, in the near-infrared region of the spectrum, shows the clearest distinction between land and water; TM Band 5, in the mid-infrared region of the spectrum, shows the moisture content of plants. This helps to detection plant water loss and early plant distress; and TM Band 7, also in the mid-infrared region of the spectrum, is particularly useful for detecting rock structures and mineral deposits.

TM Thermal Band 6
In TM Band 6, the amount of heat emitted from objects on the ground is recorded. It is used to track environmental conditions, such as air and water pollution, and can show heat levels of volcanic eruptions and heat loss from buildings and homes.

Composite Images

When Landsat images are purchased by a consumer such as an analyst, they are delivered in these seven different bands. Remote sensing analysts combine any three of the seven bands to make a color composite image. Each band is assigned one of the primary colors: blue, green, and red. Different combinations of bands show certain landscape features more clearly. The 321 visible band sequence shows a true color image of how the Earth appears to the human eye. By studying other band combinations including bands from both visible and the invisible portion (i.e. 234 combination), we can show details that we could not have detected before. Combinations can be made that can help reveal the mineral content of rocks, the moisture of soil, the health of vegetation, the physical

Landsat 7 Bands 321, Wayne County, Michigan 09/30/04

Landsat 7 Bands 742, Wayne County, Michigan 09/30/04

composition of buildings, and thousands of other invisible details. TM bands 7,4 and 2, is good for detecting different land uses. It is close to natural color, and it shows light reflected in the mid- and near-infrared regions of the spectrum. Plants are true green color, urban areas are pink or purple, grasslands are yellow or light green, and forested areas are deeper tones of green. Other common combinations are bands 1,2 and 3, which is good for mapping water sediment patterns; 2,3 and 4, and 3,4 and 7, which are both good for mapping urban features and plant types; and 4,3,2 which is especially good for mapping vegetation vigor.

Other Electromagnetic Spectrum Waves and Sources

This lesson has focused on multispectral imagery taken from space towards the Earth. There are, however, many sources of multispectral imagery. There are sensors on Earth that take images of space and interpret the various items and anomalies that occur miles above the Earth's surface. In order to see something that we cannot normally see, we can use another form multispectral imagery, X-rays (or X-radiations). Microwaves are also included on the electromagnetic spectrum.

Conclusion

Remote sensing images can include images taken from miles above the Earth. These satellites and sensors collect energy that is reflected back from the Earth's surface. Reflectance measurements and digital images created from them provide an extremely accurate representation of what surface features and objects on the ground look like. But perhaps even more importantly, digital images can show more than simply those spatial details. Satellite sensors gather image data from different portions of the electromagnetic spectrum – both visible and invisible bands. The Landsat 7 satellite captures imagery that shows small scale patterns

---

of landforms and land use over large areas at relatively low resolution. Another satellite, Quickbird, is great for showing landscape detail of smaller areas at very high resolution. There are other remote sensing satellites, both government-supported and commercially owned, that provide a range of satellite imagery products. Each of these image sources produces a specific product that satisfies a particular need.

Lesson 3: Remote Sensing & Satellite Imagery Exercise

Satellites have been collecting remotely sensed data for over 40 years. Initially many were used for defense and weather purposes, but now there are many other uses for the variety of data collected by these instruments.

Your Mission: Draw a line to connect the Band number to its common use.
1. – 7.

Band number:	Common Band Use:
Band 1	Detects rock structures and mineral deposits
Band 2	Records heat emitted from objects
Band 3	Shows clear distinction between land and water
Band 4	Shows moisture content of plants
Band 5	Shows snow cover and distinction between soil and plants
Band 6	Shows water depth
Band 7	Shows water turbidity and plant life

Landsat's Enhanced Thematic Mapper collects multispectral data along the electromagnetic spectrum.

8.   Which 3 layers would be best for detecting urban features? _____

9.   Combining TM bands 7,4, and 2, is good for detecting _____.

10.   This combination shows the Earth in true color. _____

11.   If your study included patterns of water sediment, the bands you would use would most likely be _____.

Composite Images
Landsat images are delivered in these seven different bands – all in black & white - each capturing one feature better than the other. GIS analysts study three layers at a time with red, green, and blue colors to both enhance their perception of the area and also to analyze their study area. View the images below and use them to answer the questions that follow.

Scenario: You work for the Denton County Developer's office. Holesome Golf, a company that builds golf courses, is inquiring if there is an area suitable for a new course they want to build in your area. The proposed area has been circled for you. Holesome Golf would like to find a location that has a small amount of water and not many trees for their course.

321 Composite of Denton County, Texas

12. Looking at the 321 image, without looking at the other images, what do the darker areas in the southeastern portion of the circle represent?

_____

432 Composite of Denton County,

13. In this 432 composite, what colors are assigned to each of the bands?

Band 4 = _____
Band 3 = _____
Band 2 = _____

14. The spots that were dark in the 321 are now red, what do they most likely show?

_____

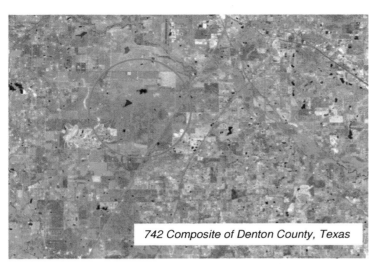

742 Composite of Denton County, Texas

15. In this 742 composite, what does the light green represent?

_____

16. What does the blue layer represent?

_____

17. Based on your analysis and their requests, does this look like an area that they will be happy to build a golf course on? Why or why not?

_____

_____

_____

Lesson 3: Remote Sensing & Satellite Imagery Lesson Review

Key Terms
Use the lesson or glossary provided in the back of the book to define each of the following terms.

1.   Electromagnetic Spectrum

2.   True Color Image

3.   False Color Image

Global Concepts
Use the information from the lesson to answer the following questions. Use complete sentences for your answers.

4.   What were satellites first used for?

5.   Explain how satellites collect data.

6.   How long have man-made satellites been orbiting the Earth?

7.   What type of information do satellites (their sensors) collect?

Sources of Satellite Imagery

Place the letter of the source next to the correct description. Some may be used more than once.

Description	Source
_____8. This program was started to support their country's national economy	a. QuickBird
_____9. This program was collaboration between NASA and the USGS.	b. ASTER
_____10. Forecasters use the info obtained from these satellites to track storms.	c. GOES
_____11. Instrument that collects info about land surface temperature.	d. SPOT
_____12. There have been 7 of these launched to this date.	e. IRS Satellite
_____13. The acronym for this satellite stands for Systeme Pour l'observation de la Terre.	f. Landsat
_____14. Stays in a fixed spot above the Earth's surface.	
_____15. Capable of delivering data with less than one meter resolution.	
_____16. Circles the globe every 26 days	
_____17. Collects data along the electromagnetic spectrum in 7 bands.	

Let's Talk About It...

Answer the following question and share the responses with your instructor and classmates.

21.   Why would an analyst use composite images instead of traditional color aerial imagery?

Lesson 4:
Identifying Features Using Imagery

One of the many benefits of using imagery in a GIS is to identify features. Both aerial photography and satellite imagery have their benefits and can be used for a variety of different applications. In Lesson 1, aerial photography was shown to be beneficial for identifying and displaying smaller features in greater detail usually requiring smaller areas of interest. Satellite imagery can be used to identify and visualize larger features in larger geographic areas. The project manager will need to be able to select the appropriate type of imagery that is best for the project. The decision to use either source of imagery depends upon the application and needs of the project.

Identifying Features

Earlier in this unit, you identified features in a true color photo of an area that was familiar. If one of the tasks in a project was simply to identify small features for areas that are familiar, true color imagery – via aerial photos or satellite imagery – may be all you need to fulfill your task. In some projects, however, identifying features may require the use of several types of imagery to efficiently locate features of interest.

For instance, what if a project manager's area of interest is an area that is unfamiliar? Whether you are familiar with the area or not, after orienting yourself to a high resolution image, you can usually identify many features that are common to other populated areas such as roads,

buildings, houses, athletic facilities, and airports. Because aerial photos are most useful for a smaller geographic area, they are good for making large scale maps of an area.

If your study area needed to include an entire county or state, aerial images would most likely not be the most efficient method of displaying image data. In most counties, it would take numerous aerial images of the county to form one complete image of the county. Due to the file size of high resolution imagery, aerial photos may cause slow performance and will require large amounts of storage space. In this

QuickBird can supply images with a .6 meter resolution. This is an area in Denton County, Texas. You can see a great amount of detail including cars streets.

Spatial Technology And Remote Sensing

STARS
Remote Sensing Education

SPACESTARS
Laboratory for GIS/Remote Sensing Education

case, a low resolution image like a **satellite image** might be more appropriate for looking at features. If analysts were familiar with the area, they would be able to identify where larger bodies of water were located, where a highly populated area is, and more. For instance, **Landsat** satellite imagery is delivered at 30 meter resolution. This is beneficial if you need to see a large amount of geographic area but is not intended for zooming in closely to see a feature. As resolution improves, more and more features can be recognized and identified. **ASTER** satellite images can be ordered to 15 meter resolution and are used for a variety of applications. Products are available at an even higher resolution image from a satellite up to .6 meters. These images can look as clear as a large scale aerial photograph taken at low altitude. One of the decisions project managers will be faced with is determining the resolution quality required to successfully complete a project.

What if your study required that you find features that were not as easily seen as looking for a football stadium or an airport? Some features' boundaries are not as easily seen or conveniently shaped. Think of a scenario where an analyst might define an accurate boundary of an urban area for identifying urban sprawl. Regardless of their familiarity of the area, it may be difficult to efficiently define all urban areas. **Multispectral imagery** can be a valuable piece of data in this situation. Multispectral imagery and false color imagery can be used to enhance the ability to see features that may not be detectable, or as obvious, as using true color imagery. Take a look at the images below, notice how the 742 image shows urban areas in more detail.

These are both images of a western portion of Harris County, Texas. The 321combination on the left shows a separation between grassland and urban areas as does the 742 on the right.

What Should a Project Manager Choose When Selecting Imagery?

Project managers have a lot to consider when working through a project. One of the many duties they may have is to acquire imagery in a previously unstudied area. In acquiring imagery, they need to focus on purchasing not so much what is the "best looking" or all image(s) of an area (unless that is the goal of the project) but focus on acquiring the image(s) that will best fulfill their project's needs.

There are numerous advantages and disadvantages of obtaining **high resolution** imagery. Specifically, some advantages, mentioned above, are that high resolution is better suited to see smaller features and shows much more detail than **low resolution** images. On the other hand, there are drawbacks of obtaining high resolution imagery. The general rule with any type of imagery is the better the resolution, the higher the price. Also if the image has to be **tasked** (meaning the image has not been acquired by any organization), the cost will be more than if you can find an image that has already been taken. High resolution images take up a lot of memory. If you have to load several on your system at one time, it can lock up your system. The file sizes of the images are also large. A decision to use many high resolution images when a single low resolution image is sufficient may contribute to additional costs in storing or transporting the data.

Just as there are several advantages and disadvantages of high resolution, there are also pros and cons for low resolution too. Some advantages are that low resolution are better for viewing a larger geographic area and is ideal for small scale maps because of their smaller file size and lower memory requirements. The imagery is usually easy to acquire with less cost. The disadvantages are that low resolution is not good for seeing small features on a map and therefore would not be useful for large scale maps. The key for the project manager is purchase or find what the appropriate image they need for the job.

Conclusion

Geographic features can be identified using satellite imagery. Using high resolution imagery allows the user to identify smaller features and is good for creating large scale maps. These images are usually obtained from aircraft or satellites that are on a low orbit cycle. High resolution satellite imagery can appear similar to aerial photography in the amount of detail that can be seen and how easily it is to identify geographic features.

Low resolution imagery, on the other hand, is ideal for studying larger geographic areas and its features. Multispectral low resolution imagery can be used to study a wider area such as a county or even a state.

When selecting imagery for a project the project manager must keep in mind not only potential cost of obtaining an image, but also what the actual imagery needs are for the project. If the project requires identifying large features or boundaries, then low resolution imagery may be all that is needed. If the project requires identifying small features like buildings or parks, then a higher resolution may be needed.

Lesson 4: Washington, DC Exercise Worksheet Assessment

Directions: Use the Quickbird natural color satellite image of Washington, D.C. on the poster, and the Washington, D.C. map located in Appendix A, to locate as many landmarks as you can and note the location of each in the grid (some may span more than one grid).

Observe how these same landmarks are displayed in the lower resolution ASTER color visible infrared image of the same area. Note on the worksheet any significant differences in their appearance.

	Landmark Name	Grid Location Quickbird	ASTER	ASTER Description
1.	*Bureau of Engraving & Printing*	*C4*	*H4*	*blue color, no building detail*
2.	Convention Center			
3.	The Ellipse			
4.	Jefferson Memorial			
5.	The Mall			
6.	Marina Piers			
7.	Museum of American History			
8.	National Archives (Nat. Arch.)			
9.	National Gallery of Art			
10.	Natural History Museum			
11.	Reflecting Pool			
12.	Ronald Reagan National Airport			
13.	Tidal Basin			
14.	Holocaust Memorial Museum			
15.	Vietnam War Memorial			
16.	Washington Monument			
17.	West Potomac Park			
18.	White House			

Directions: Use your PowerPoint notes and your manual to answer the following questions.

19. You work for a swimming pool company who wants to locate houses in a specific neighborhood without pools to market their pools to. Which type of imagery would be best to use? Why?

---

20. You need to quickly acquire an image of your county that only needs to be clear enough to tell where the larger bodies of water are. Which type of imagery would be best to use? Why?

21. You are assigned to a county that you are unfamiliar with to find areas where the highest density of trees is. Which type of imagery would be best to use? Why?

22. With advantages like being able to see small features in detail (even lines in parking lots), high resolution imagery can help in easy identification of features. What are some of the disadvantages of using high resolution imagery?

23. Low resolution imagery, like Landsat imagery, is typically not used for seeing small features. What are some advantages for using it?

24. When selecting imagery for a project, what should the project manager's focus be?

Lesson 4: Identifying Features Using Imagery Lesson Review

Key Terms
Use the lesson or glossary provided in the back of the book to define each of the following terms.

1. Satellite Imagery

2. Multispectral imagery

3. Tasked

4. High resolution imagery

5. Low resolution imagery

Global Concepts
Use the information from the lesson to answer the following questions. Use complete sentences for your answers.

6. If the area of interest for a geospatial study included an entire state, why would it not be advisable to use aerial photographs?

7. List two pros and two cons for obtaining high resolution imagery.

8. List two pros and two cons for obtaining low resolution imagery.

Let's Talk About It...
Answer the following question and share the responses with your instructor and classmates.

9.   If you had to study an entire county to determine areas that were experiencing drought conditions, which type of imagery would it be best to use? Why?

Introduction to GPS Technology

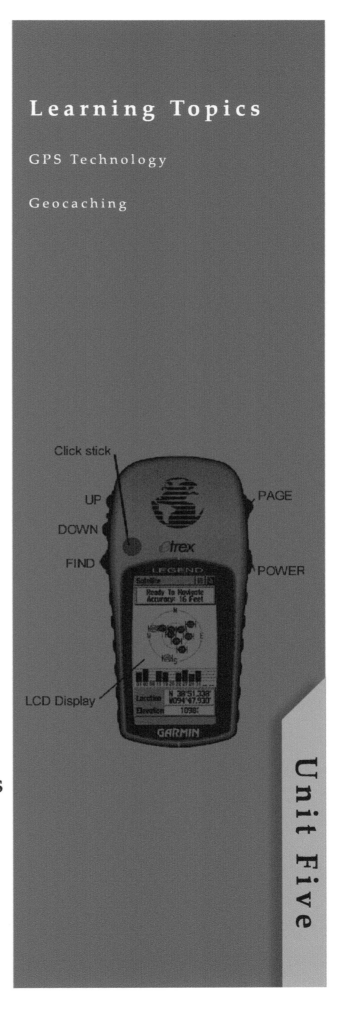

Click stick

UP

DOWN

FIND

PAGE

POWER

LCD Display

Introduction to
Geographic Information Systems and Remote Sensing
Concepts

Unit Five

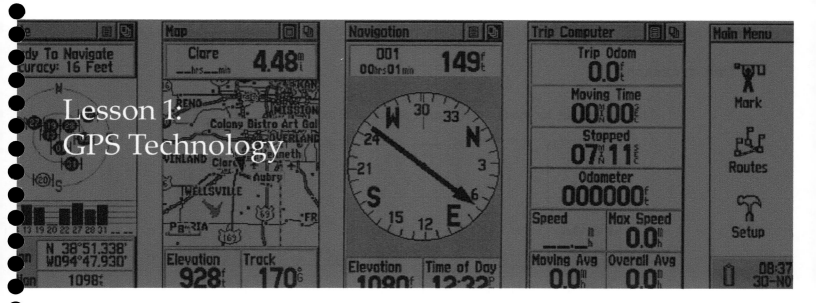

Lesson 1: GPS Technology

As you already know from the information in the lesson and the PowerPoint presentation, GPS technology can enhance GIS analysis by collecting data recording the geographic location of features in a study area. In this exercise, you will use a GPS unit to determine your geographic position on the surface of the Earth.

Please note that this lesson outlines steps for collecting GPS data using a Garmin eTrex Legend unit. Please take the time to read through these instructions to ensure the proper functioning of the GPS unit. Steps required to collect data using other units will differ from these instructions.

Examining the GPS Unit:

Each group of 2-4 students will use a Garmin eTrex Legend GPS unit.

Prior to using the GPS unit, read the user manual that was included in the packaging for complete information on use and care of the unit. For this lesson, you will need to familiarize yourself with the following unit features:

In order to get the best satellite signal possible, you should have a clear view of the sky and hold the unit flat and away from your body. Use the following instructions to use the GPS unit:

1. **Turn on** the unit by pressing the **POWER** button on the side of the unit.

The main pages for the Garmin eTrex Legend GPS include:

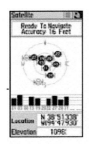   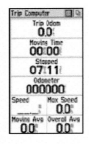 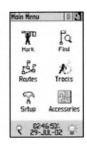

Satellite Page   Map Page   Navigation Page   Trip Computer   Main Menu

2. When the unit is initially turned on, the **Satellite Page** appears as the unit begins receiving signals from satellites within its range. The **Satellite Page** provides information on satellite tracking, informs you when the unit is ready and shows your location coordinates. There is a number assigned to each of the satellites in orbit. The satellites within the reception range of your GPS unit show up on the screen in their positions in the sky. The center point of the circle represents the point directly above your head. As you move out away from the center to the outside rings, the satellites are positioned towards the horizon. The satellites sitting on the outside circle are orbiting on the horizon, and your GPS unit may have difficulty receiving a strong signal from them. The bars at the bottom of the screen show the strength of the signal received from each satellite. The GPS unit will not display a longitude/latitude position until it has received a strong enough signal from at least **4** satellites. You will get a message on the screen once a position is determined.

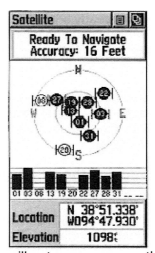

You can use the **Thumb Stick** to highlight options on the screen from the **Options Menu** button. Selections listed in the **Satellite Page Options Menu** include:

Use with GPS Off – You should use this option when you are indoors or do not have an unobstructed view of the sky in order to save battery power.
Track Up/North Up – This option allows you to specify the satellite skyview as Track Up (the direction of travel) or North Up (skyview oriented north).
New Elevation – You can manually enter the known elevation to increase accuracy.
New Location – This option should be used if you travel more than 600 miles from where you last used the unit.

You can use the **Page** button on the side of the unit to move through each of the main pages.

Knowledge Knugget

To highlight and select an item from the **Options Menu** at the top right corner of a page, use the **Thumb Stick** to highlight the menu button and then press the **Thumb Stick** to open the menu. Highlight the desired option and then press the **Thumb Stick** again to launch the option. You can also use the Thumb Stick to highlight the **Main Page Menu button** and then select the main page that you want to view with the **Thumb Stick**.

---

3. *Press* the **Page button** until you advance to the **Map Page** OR *use* the **Thumb Stick** to highlight the **Main Page Menu button** and then *select* the **Map Page**. The black arrow on this page displays your location within the context of a map of your community. You can use the **Up** (**Zoom In**) and **Down** (**Zoom Out**) buttons on the side of your unit to zoom in and out on the display.

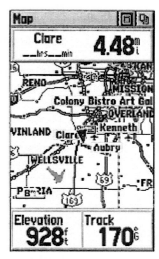

You can use the **Thumb Stick** to highlight options on the screen from the **Options Menu** button . Selections listed in the **Map Page Options Menu** include:

Pan Map – This option allows you to move the map pointer about the area on the map.
Stop Navigation – This option stops the navigation to a destination.
Hide Nav Status/Show Nav Status – This option shows or hides the navigation status window.
Hide Data Fields/Show Data Fields – This option shows or hides the data fields at the bottom of the Map Page screen.
Setup Map – This option allows you to edit map elements such as changing text size, map orientation and detail.
Measure Distance –This option allows you to determine distance from one point to another on the map.
Restore Defaults –This option restores the settings for the Map Page to its original factory settings.

4. *Press* the **Page button** until you advance to the **Navigation Page** OR *use* the **Thumb Stick** to highlight the **Main Page Menu button** and then *select* the **Navigation Page**. . As you move, this page shows you the direction of your travel, the time to the destination, and the distance to the destination.

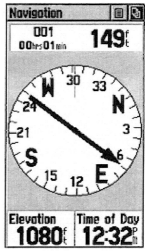

You can use the **Thumb Stick** to highlight options on the screen from the **Options Menu** button . Selections listed in the **Navigation Page Options Menu** include:

Stop Navigation – This option stops the navigation to a destination. It is grayed out if you are not navigating.
Bearing Pointer/Course Pointer – This option allows you to toggle between Bearing (pointer directed at the destination) and Course (shows amount of correction needed to be on course).
Big Numbers/Big Compass – This option allows you to toggle between the sizes of the compass and the data fields at the bottom of the screen.
Restore Defaults – This option restores the settings for the Navigation Page to its original factory settings.

5. *Press* the **Page button** until you advance to the **Trip Computer Page** OR *use* the **Thumb Stick** to highlight the **Main Page Menu button** and then *select* the **Trip Computer Page**. . This page displays eight (8) different types of selectable navigation data that display current information as you navigate. The default navigation data

settings are shown in the graphic on the following page. This page is customizable depending on your navigation needs.

You can use the **Thumb Stick** to highlight options on the screen from the **Options Menu** button ▣. Selections listed in the **Trip Computer Page Options Menu** include:

Reset – This option allows you to clear old data when you start a new trip.
Big Numbers – This option allows you to change the sizes of the numbers for the various data displays contained on this page.
Restore Defaults – This option restores the settings for the Trip Computer Page to its original factory settings.

To specify the data fields that you want to include on the Trip Computer page, *use* the **Thumb Stick** to highlight the field then *press* the **Thumb Stick** to open the **Data Field Options** menu. Using the **Thumb Stick** and the **Up** or **Down** buttons on the side of the unit, move to the desired data option and *press* the **Thumb Stick** to insert that data option in the data field.

6. *Press* the **Page button** until you advance to the **Main Menu page** OR *use* the **Thumb Stick** to highlight the **Main Page Menu** **button** 🔲 and then *select* the **Main Menu Page**.

The **Main Menu page** appears differently from the other main pages that have been previously examined. The **Main Menu page** contains icons that each opens a separate page within this page. These Main Menu pages can be accessed using the **Thumb Stick** and are as follows:

Main Menu Pages

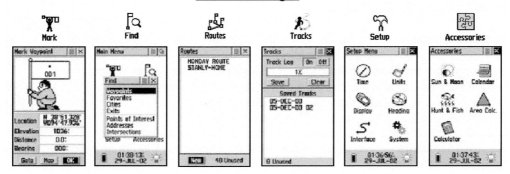

These pages are used as follows:

Mark – This page allows you to mark and store a waypoint for your current location that can be marked on the Map Page and eventually downloaded or managed for use in a GIS project.

Find – This page allows you to find waypoints, cities, exits, points of interest, addresses or other data as part of a route.

Routes – This page allows you to create routes and store them for future use.

Tracks – This page allows you to access the track log and any tracks you may have saved.

Setup – This page allows you to specify the format of items such as the time, units of measure, etc.

Accessories – This page gives you access to various tools including a calculator, an area calculator, a calendar, sun and moon data, and fishing and hunting data.

Preparing the Unit for Data Collection:

You will now check the setup formatting for a number of items including the units of measure for the positions you will collect and the time of day.

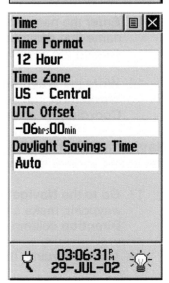

7. If you are not still viewing the Main Menu page, ***press*** the **Page** button on the side of the unit until you get to this page.

8. ***Move*** the **Thumb Stick** to ***highlight*** the **Setup Page** and ***press*** the **Thumb Stick** to open it.

9. ***Highlight*** the **Time** icon and ***open*** this option.

10. ***Confirm*** the settings for time and make any changes necessary for your particular time zone using the **Thumb Stick**. Note that you can use the 12-hour AM/PM format or the 24-hour military format. Also note that the **UTC Offset** (number of hours of difference in Greenwich Mean Time) is entered for you based on the **Time Zone** that you specify. Check the time setting at the bottom of the page to make sure that the settings that you entered are correct.

When you have finished making changes or checking these settings, ***close*** this page by ***highlighting*** the **Close** button ▣ at the top right corner of the page and ***pressing*** the

Thumb Stick OR *press* the **Page** button until you get back to the **Main Menu** page.

11. In the **Main Menu** page, *highlight* the **Units** icon and *open* this option.

12. *Toggle* to the **Position Format** setting and *press* the **Thumb Stick** to view the options for this setting. *Set* this option as *hddd.ddddd°*.

    For **Map Datum**, open the options for this setting. (Remember your lessons about Map Datums from Unit 2?) Look at the various datums that are listed. The default datum is World Geographic System 1984 (WGS 84). Keep the datum set as the default.

13. Familiarize yourself with the other unit settings listed and the various options included for each.

    When you have finished making changes or checking these settings, *close* this page by *highlighting* the **Close** button

    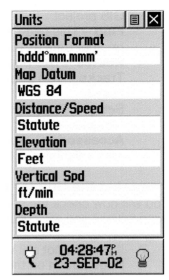 at the top right corner of the page and *pressing* the **Thumb Stick**.

Collecting Data using GPS:

You will now collect waypoint data with the GPS unit. Proceed to the position where you want to begin determining position. For this exercise, choose a landmark such as a bench, a sign, etc.

14. With your GPS unit on the **Main Menu page**, *press* and *hold* the **Thumb Stick**. The **Mark Waypoint** page will appear. *Highlight* the waypoint number in the flag on the page and *click* to select it. An alphabetic and numeric keypad screen will appear to allow you to name your first waypoint.

15. *Enter* the name for your first waypoint by using the **Thumb Stick**. Give this waypoint a name that describes its landmark such as Bench, Sign, etc. When you have finished naming the waypoint, *click* **OK**.

16. *Write* the **Waypoint 1** name in the **Waypoint Name** column on the **S1U4L5 GPS Worksheet** that accompanies this lesson. *Write* the location coordinates in the **Coordinates** column on the **GPS Worksheet** for that waypoint.

Navigating using GPS:

Determine another landmark on your campus to record as a second waypoint.

17. *Go to* the **Navigation** page on your GPS unit. As you are walking to your second waypoint, make a note of the direction you are traveling. *Write* this direction in the **Direction** column on the **GPS Worksheet** that accompanies this lesson.

18. *Mark* and *name* the waypoint for the second landmark you have chosen.

---

19. ***Choose*** two more landmarks as waypoints. ***Mark*** and ***name*** each of them. Be sure to ***record*** the **name, coordinates** and **direction** of each of the waypoints.

Finding Waypoints:

You may need to go back and look at the waypoints you have collected in case you didn't record a waypoint name or coordinates. You can also use this opportunity to check the record that you have made on the **GPS Worksheet**.

20. When you have finished collecting the waypoints, ***press*** the **Page** button to return to the **Main Menu** page.

21. Use the **Thumb Stick** to ***highlight*** the **Find** icon and ***press*** the **Thumb Stick** to open the **Find** page.

22. ***Select*** **Waypoint** from the **Find** list. ***Select*** to find the waypoint **By Name**.

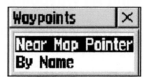

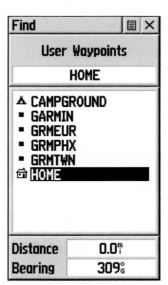

23. Using the **Thumb Stick**, ***select*** the desired waypoint to display its information page.

Returning to a Waypoint:
At some point, you may need to go back to the actual location of a waypoint that you marked.

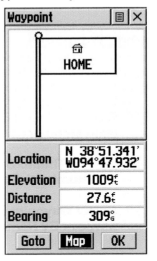

24. Once you display the waypoint information that you want to review, use the **Thumb Stick** to *click* **Goto**. The **Navigation Page** will appear and will direct you to the waypoint.

    Use the **directional arrow** and the **distance reading** at the top of the page to navigate to the waypoint position.

Viewing All Waypoints:

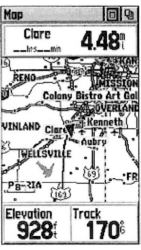

25. When you have finished navigating to the waypoint, *press* the **Page** button to go back to the **Map** page.

26. *Press* the **Up** (**Zoom In**) and **Down** (**Zoom Out**) buttons on the side of the GPS unit to view the waypoints on the map.

Measuring Distance :
Your GPS unit can provide you with distance measurements based on the waypoints that you collected.

27. With the **Thumb Stick** *open* the **Options menu**. *Select* **Measure Distance**.

28. Use the **Thumb Stick** to move the reference arrow to **Waypoint 1**.

29. *Press* the **Thumb Stick** to select the waypoint.

30. *Select* **Waypoint 2** with the **Thumb Stick**. (**Do not** press the Thumb Stick this time.) The waypoint name, distance and direction from Waypoint 1 will display. *Record* the direction and distance from Waypoint 1 to Waypoint 2 in the Waypoint 2 record on the **GPS Worksheet** that accompanies this lesson.

31. Determine the distance from Waypoint 2 to Waypoint 3; and Waypoint 3 to Waypoint 4. Record these distances on the **GPS Worksheet** that accompanies this lesson.

Calculating Area using GPS:

You can use the Garmin etrex Legend GPS unit to calculate area.

32. *Press* the **Page button** until the **Main Menu** page appears.

33. Use the **Thumb Stick** to *highlight* and *select* **Accessories**.

34. *Select* **Area Calculator**. The Area Calculation page will appear on your screen. (It looks similar to the Map page.)

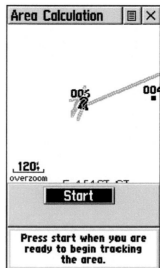

35. Determine a portion of your campus that you would like to use to perform an area calculation. For example, you could calculate the area of the front lawn, a parking lot, a common area, etc. When you have chosen the area of campus, proceed to a place on the boundary of the area. *Record* the name of the campus area that you will measure on the **GPS Worksheet** that accompanies this lesson.

36. *Press* the **Thumb Stick** to **Start**.

37. *Walk* the boundary of the area until you return to your starting point. *Press* the **Thumb Stick** to **Stop**.

    Record the area that displays on the **GPS Worksheet** that accompanies this lesson. The area calculation will be displayed with acres as the unit of measure.

Turning Off the GPS Unit:

When you have finished working with the GPS unit, turn off the unit to save battery power for future uses.

38. *Press* and *hold* the **Power** button.

39. Turn the unit in to your instructor.

Lesson 1: GPS Technology Recording Log

As you complete the activities for the GPS lesson, enter your information in the appropriate spaces:

	Waypoint Name	Coordinates	Direction	Distance
Waypoint 1			---	---
Waypoint 2				
Waypoint 3				
Waypoint 4				

Campus location used for area measurement:

Area calculated:

Lesson 1: GPS Technology Lesson Review

Key Terms
Use the lesson or glossary provided in the back of the book to define each of the following terms.

1. **GPS Constellation**
2. **Elevation**
3. **Waypoint**
4. **Ground truthing**

Global Concepts
Use the information from the lesson to answer the following questions. Use complete sentences for your answers.

5. How many satellites do GPS units need to receive signals from to have accurate readings?

6. How many (minimum) satellites do you need to obtain an altitude (z) reading?

7. Determine if the following statement is true or false and explain your answer. *All GPS units can only collect point data.*

Use the notes from the PowerPoint presentation to answer the following questions.
8. Using the Satellite page, what do the black and gray bars at the bottom represent?

9. What procedures should you follow before using the unit for the first time?

10. Besides providing you with the direction that you are moving and the time of day, what two other pieces of information does the Navigation page show you?

11. If you want to know how far you have traveled and how fast you have traveled there, which page would you refer to?

12. What does the Mark Waypoint page allow you to do?

Let's Talk About It...

Answer the following question and share the responses with your instructor and classmates.

13. What are some ways in which people have used GPS technology in their lives?

Lesson 2: Geocaching

	A	B	C
1	Waypoint	Longitude	Latitude
2	1	-93.48506	35.465
3	2	-93.48506	35.46501
4	3	-93.48507	35.46501
5	4	-93.48511	35.46501
6	5	-93.48513	35.46502
7	6	-93.48515	35.46502
8	7	-93.48517	35.46503
9	8	-93.4852	35.46505
10	9	-93.48523	35.4651
11	10	93.48523	35.4651

When the Global Positioning System was initially developed in 1978, its primary purpose was for national security. As a result of this application, the scrambled signals that any nonmilitary receivers would receive were not highly accurate, only getting the user within 100 meters of a position. This of course would get GPS users like hunters and fishermen back to a certain area, but not necessarily back to a specific point. In May of 2000, the President announced the scrambling of the GPS signals, called Selective Availability, would be turned off. This now meant that GPS users could now benefit from being within 10 to 15 meters of their intended targeted area. Soon thereafter, a new outdoor GPS sport was developed called **geocaching**.

Geocaching

Geocaching is a relatively new sport that utilizes GPS technology to create a high-tech game of hide and seek. **Geocachers** hide **geocaches** (also called caches) in various locations and then post the locations on geocaching websites so that others can use GPS units to find the caches. The sizes of these caches will vary per location. They are usually composed of a substance that is waterproof and may be as small as a mint box or as large as the size of a mailbox. In some cases, the cache will be painted or "camouflaged" to blend in with its environment.

Geocaches of all shapes and sizes are hidden all over the world. The given coordinates led the user to this location. Can you spot the geocache in this picture?

Rules of the Game

The following are guidelines that are understood as geocaching etiquette.

1. Each geocache will have a logbook in it. (If the cache is large enough, a pen or pencil may also be included to use.) This logbook is provided to allow users to record proof of their visit by adding their name, date, and usually their hometown or state. It can be interesting to flip through the logbook to see where previous geocachers have traveled from. Typically, once you find a cache, you will add your name to the end of the list and replace it back in the container for the next geocacher to find.

Spatial Technology And Remote Sensing

STARS
Remote Sensing Education

SPACESTARS
Laboratory for GIS/Remote Sensing Education

2. Some geocaches will contain small items. The rule with these caches is if you take something, you must replace it with something else. When you go geocaching, take some small items with you that you are willing to trade.

3. Return the cache back to its original location. This will set the geocache up for the next person.

It is a good habit and great for the environment for geocachers to pick up any trash in the areas or parks containing geocaches. Some geocaching organizations designate dates for geocachers to meet in certain locations to have a clean up day.

Geocaching.com

The website www.geocaching.com provides valuable insight and support for this sport. This site is a clearinghouse of locations of geocache sites all over the world. Once you enter your zip code, you will be able to find a list of geocaches in your area. You will use this site in this lesson.

Conclusion

Geocaching is an outdoor activity that uses GPS technology in a high-tech hide-and-seek treasure hunt game. Geocachers can go online to find both locations of geocaches in an area and to post cache location coordinates for others to find their caches. Caches contents constantly change as geocachers take and leave items in the caches. A few simple rules of "geocaching etiquette" should be followed when participating. There are several geocaching websites that allow users to record and locate geocache locations worldwide.

The geocache from the previous image was found under a rock and leaves at the base of a fencepost. Can you think of places in your community where geocaches may be hidden?

Lesson 2: GPS Geocaching Exercise

Geocaching (pronounced geo-cashing) is a relatively new activity that is in essence a hide-and-seek treasure game for GPS users. Geocaching involves users setting up "caches" or collections of various items for the sole purpose of other users finding the items, taking items of interest and leaving items for others to find. Users who set up the caches use the Internet to share the locations of the caches. Geocache hunters then use GPS units to find the geocache locations. In this enrichment activity, you will continue to explore the use of GPS through a geocaching exercise. You will use the Geocaching.com website for this exercise, which is a very good resource for background information about this sport. You may access this site through a link from the **SPACESTARS** website **www.spacestars.org** by clicking **Links/Training Links**.

Locating Geocaches in Your Community:
Right now there are geocaches in your area. One way to find out where these are is to locate online sources that will supply coordinates for geocaches in your community.

1.   *Launch* your **Internet browser** and *navigate* to the **SPACESTARS website www.spacestars.org**.

2.   In the **Links/Training Links** section, go to the Geocaching website **www.geocaching.com**.

3.   In order to obtain coordinates for the geocaches, you must agree to the terms and conditions of it use. Do this by creating a free account by *clicking* on MY ACCOUNT at the left side of the home page screen.

4.   *Click* Create a membership! to go to the Join the Geocaching.com Community page. There are two types of memberships to select from. The Basic Membership, which is free, is all that you will need for this lesson.

5.   *Click* Get a Basic Membership and enter you information to create a new account. Once you have set up your account, you will need to validate it by going to your email account to follow the validation steps.

6.   Once you account is validated, *enter* your **zip code** in the **search** section to retrieve a list of caches in your area. *Click* Go... . A list of geocache sites in your area including the distance to the sites is returned.

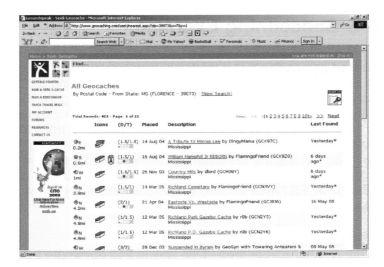

7. **Click** the link for a geocache site that interests you.

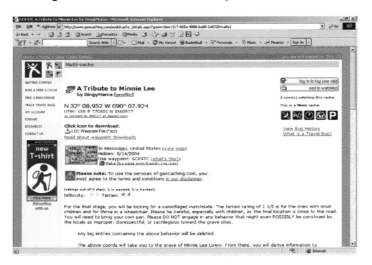

8. **Enter** the coordinates provided on the site into your GPS unit. **Name** the waypoint a name appropriate for the cache site.

9. **Use** the **Goto** option on the **Mark Waypoint** page to navigate to the cache.

   Keep in mind that on the Geocaching.com website you can report the inventory of items in geocaches that you find. The general rule is that if you take an item from a cache, you should leave something in its place. You can also record the coordinates of geocaches that you have hidden on this website.

10. When you have finished**, turn off** your GPS unit and **exit** your Internet browser.

Lesson 2: GPS Geocaching Geocaching Worksheet

As you complete the activities for the Geocaching enrichment lesson, enter your information in the appropriate spaces:

Coordinates for Cache 1 (from instructor):

Inventory of Cache 1:

Coordinates for Cache 2 (from student group):

Inventory of Cache 2:

Coordinates for Cache 3 (from student group):

Inventory of Cache 3:

Coordinates for Cache 4 (from student group):

Inventory of Cache 4:

Coordinates for Cache 5 (from student group):

Inventory of Cache 5:

Coordinates for Cache 6 (from student group):

Inventory of Cache 6:

Coordinates for Cache 7 (from student group):

Inventory of Cache 7:

Introduction to Aerospace Technology

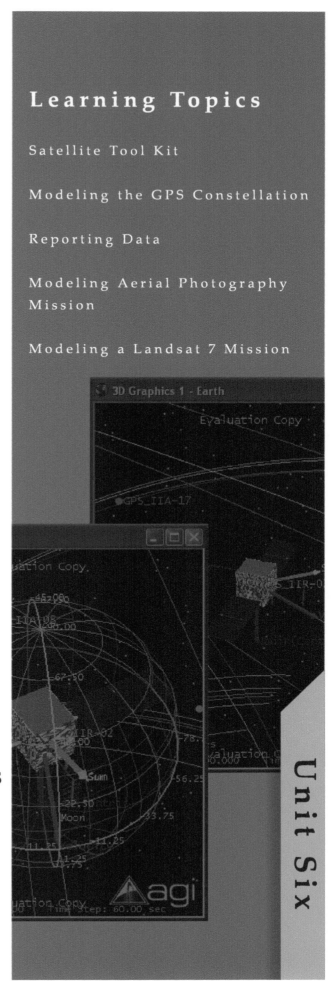

Introduction to
Geographic Information Systems and Remote Sensing
Concepts

Unit Six

Lesson 1:
Satellite Tool Kit Familiarization

In remote sensing, geographic information systems, and GPS technology, you have access to a variety of data types. Much of this data is created from sensors' observations above the earth's surface, but how are these sensors guided and directed? Aerospace technology uses highly advanced software to "map" the sensors above the Earth. Satellite Tool Kit (STK) a multifaceted software tool designed for aerospace professionals that simulates various satellite orbits and models aerospace applications. You will be interested to know that this is the actual software that NASA uses to model all of its missions including space shuttle launches, Mars probes, the International Space Station, the Hubble telescope, etc. In addition, the Defense Department uses STK to model its missile systems.

With GIS, users work toward representing the Earth two dimensionally to simplify location, direction, and measurements. When you map objects above the earth's surface that have varying altitudes and travel at high speeds, you must looks at the world in three dimensions. How exactly do the altitudes vary? There are a few basic ways that a satellite can travel around the Earth. In a **Low Earth Orbit (LEO)**, a satellite with a relatively short period (or orbits the Earth in 90 minutes) orbits the Earth at a height of less than 2,000 km above Earth's surface. A **Medium Earth Orbit** has an orbit up to 10,000 km.

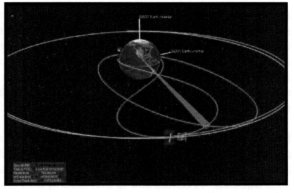

Some satellites travel in circles, maintaining a close to constant altitude while traveling around the Earth. Some satellites very and travel in an elliptical orbit, meaning that they may pass close to the Earth and then travel to higher altitudes. In the image above we see two satellite paths that move around the Earth, but vary in their altitudes. The maximum point of distance from the Earth is called the **apogee**, while the minimum point is the **perigee.** The third type of orbit is the **Highly Elliptical Orbit**, which may travel up to a 50,000 km apogee and have perigee as close as 500 km.

In the previously described orbits, satellites are traveling faster than the Earth so to speak. In some situations a satellite may need to view the same portion of Earth for an extended period of time. In a **Geostationary orbit**, a satellite may orbit over a fixed position over the Earth. This can be useful in observing weather and holding extended periods of communication over one area.

The above orbits are in part based on the function of a satellite.   How much does the satellite need to view? The **swath** is the area of coverage for a satellite's sensor. With a low orbit the satellite has a smaller swath. In higher orbits, the sensor has a larger swath...similar to the view of the Earth you have in an airplane versus on top of a small building. High Resolution imagery satellites usually have smaller swaths and are closer to the Earth for better quality imagery. By contrast, GPS satellites affect large groups of people simultaneously and have travel in higher orbits for a larger swath for maximum coverage. The same is true for telecommunication satellites.

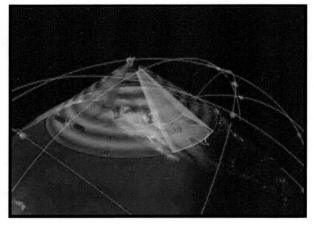

What other considerations are there when traveling above and over the Earth? In the medium earth orbit definition, the satellite had a "short period" of 90 minutes or 90 minutes per revolution. In this case the satellite will actually pass out of the sun's range into "night" or a shadow of the Earth. When the satellite is in a shadow and completely out of sight of the sun, the satellite is in **umbra.** When it is in shadow and partially in view of the sun it is said to be in **penumbra**. The satellite's capabilities may change when in umbra or penumbra. For instance, a remote sensing satellite that takes natural color imagery is not useful without the sunlight illuminating the Earth's surface.

Another factor in Satellite modeling is time. TIme is a large factor in modeling and planning satellite paths. There are many satellites orbiting and many of these satellites need to make observations at a certain time. While working in STK, the **time period** is the total duration of the analysis. The **epoch** is the fixed instance that an event will occur that serves as a reference for all other times in the scenario. For instance, in a scenario moving a satellite to make an observation, the epoch may be the instance that an image is taken. All planning for the satellite must include activities that happen before and after that.

Before looking at the software, users need to know one last essential. Above concepts focused on satellite interaction with the Earth. Often satellites are used to communicate with facilities and receivers on the ground. These ground based objects have **constraints** and factors that limit **access**. A constraint is something that can limit the access that a facility or target has to a satellite sensor. That may be a tree or a building or anything else that could potentially obscure contact with a satellite.   These constraints limit access or the time that one object can see or access another object.

When working with STK or in the satellite industry, knowledge of software is crucial, as is understanding the importance of planning the scenarios and processes that take place. STK like ArcGIS has the goal of communicating. One of the major requirements in communicating is using industry specific terminology. The above terminology can help you adjust to industry terminology and ideas.

When applying this knowledge in STK, users have four basic objects to consider. Satellites. Facilities. Targets. Sensors. **Satellites** are vehicles that are placed in Earth's orbit to view and collect physical and human data about the Earth. **Facilities** are objects that are defined as stationary locations on the Earth's surface. They may represent ground stations, launch sites, tracking stations, or other structures providing satellite support. **Targets** are locations that are identified for observation by the satellite sensor like the coordinates that may be the location of an area to be imaged. **Sensors** are equipment such as optical or radar sensors, receiving or transmitting antennas, or lasers; sensors are subordinate to (sub-objects of) the object to which they are attached.

Lesson 1: STK Familiarization Exercise

Satellite Tool Kit (STK) is a multifaceted tool designed for aerospace professionals. This exercise demonstrates its many powerful features, along with its dynamic flexibility of use. Here, you will begin to walk through some of the many capabilities of **STK**: creating objects, determining access and creating output, such as reports and graphs.

In this lesson you will:
- Create satellites and define their basic and graphics properties
- Generate satellite accesses, reports, swaths and display lighting conditions
- Define facility access and constraints
- Define sensor attributes and analyze sensor access
- Create a report, graph, dynamic display and strip chart

1. To *launch* the **Satellite Tool Kit**, *click*

    on the **STK 9 Icon** .

    You will see the HTML viewer.

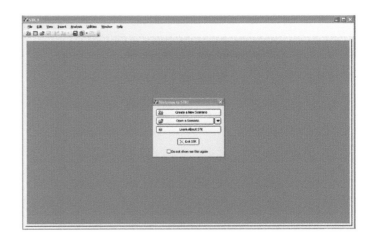

2. *Select*

    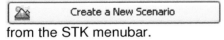

    from the STK menubar.

3. **Name** the scenario **S1U6L1_XX** (where XX is your initials). Ensure that the Location is your student directory.

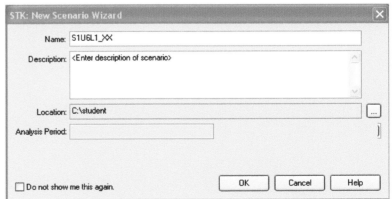

---

4. The **2D** and **3D Graphics Windows** open. Close the **Insert STK objects** window.

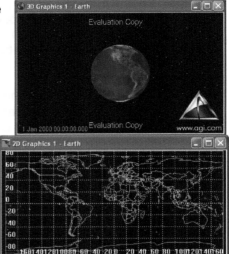

Creating a Basic Scenario

1. In the **Object Browser**, *right click* on 🖼 **S1U6L1_XX** and *select* **Rename**. Rename the scenario **Basics**. *Right Click* 🖼 **Basics** and *select* 🖹 Properties Browser .

   Here you will work with basic properties of the flight of your satellite, including time, animation, and units.

2. In the **Time Period Menu**, *confirm* the following **Analysis Period** options:

Field	Value
Start Time	1 Jan 2000 00:00:00.00
Stop Time	2 Jan 2000 00:00:00.00
Epoch	1 Jan 2000 00:00:00.00

   Note: *Each scenario has a time period and an epoch. The time period defines the general time span for analysis. The epoch serves as a reference for all other times in the scenario.*

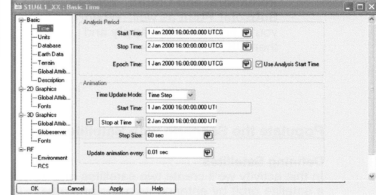

3. In the **Animation section**, *select* the following options, if not already set:

Field	Value
Start Time	1 Jan 2000 00:00:00.00
Loop At Time	2 Jan 2000 00:00:00.00
Time Step	60 sec
High Speed	

4. In the **Units menu**, *confirm* the following options:

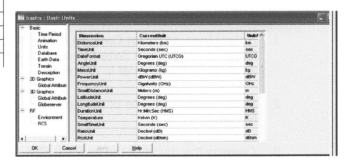

Units	Change Unit Value
Distance Unit	Kilometer (km)
Time Unit	Seconds (sec)
Date Format	Gregorian UTC (UTCG)
Angle Unit	Degrees (deg)
Mass Unit	Kilograms (kg)

5. *Select* the **Description Menu** and, in the **Short Description** field, *type* **My Basics Scenario**. When you finish, *click* **OK**.

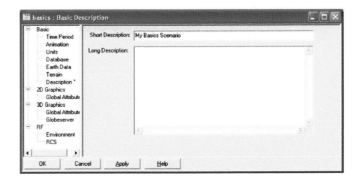

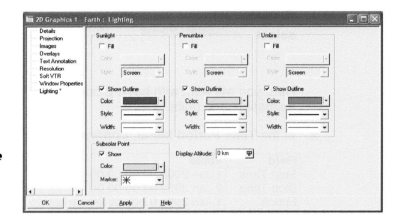

6. With the **2D Graphics Window** highlighted, *click* the **Map Graphics Properties** button ⬚ on the Toolbar

7. In the **Lighting Menu**, *enable* **Show Oultine for Sunlight, Penumbra, and Umbra.** *Enable* **Subsolar Point** as well. When you're finished *click* **Apply** and then **OK**.

Populate the Scenario with Satellites

Defining Satellites

In this activity we'll create two satellites and work with their graphic properties. First, we'll create a satellite orbit by entering its basic properties. Then we'll use the Orbit Wizard to propagate a satellite orbit and take a look at some different Graphics Properties. We'll look at different line styles, visible sides of the orbit, ground track lead types, lighting styles and swath options.

1. *Select* **Insert/Object Catalog** from the toolbar to *open* the **Object Catalog.** *Highlight* **Satellite** and *click* **Insert**. If the **Orbit Wizard** appears, *click* **Cancel** to dismiss it.

2. *Right click* the name of the new satellite  and *select* **Rename**.

3. *Change* the name of the new satellite to **leo** ✷ leo (which stands for **low earth orbit**).

4. *Right click* ✷ leo to open the ▦ Properties Browser , and *enter* the following values into the **Orbit tab:**

Field	Value
Propagator	J4 Perturbation
Start Time	1 Jan 2000 00:00:00.00
Stop Time	2 Jan 2000 00:00:00.00
Step Size	60 sec
Change **Semimajor** Axis to **Apogee Altitude**	600 km
Perigee Altitude	600 km
Inclinat	75 deg
Argument of Perigee	0.0 deg
Change RAAN to Lon. Ascn. Node	10 deg
True Anomaly	0 deg

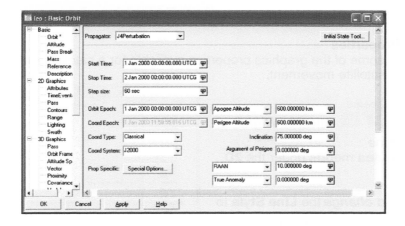

When you finish, ***click***   OK   .

5. ***Insert*** another **satellite**, as you did before.

6. When the **Orbit Wizard** opens, ***click*** Next to go to the second screen, ***select*** **Repeating Ground Trace** and ***click*** Next again.

7. In the third window of the **Orbit Wizard**, ***set*** the following options:

Option	Description
Approximate Revs Per Day	4
Inclination	45 deg
Number of Revs to Repeat	4
Longitude of First Ascending Node	5 deg

8. ***Click*** Next again and ***set*** the following options:

Option	Description
Orbit Start	1 Jan 2000 00:00:00:00
Orbit Stop	2 Jan 2000 00:00:00:00
Time Step	60.00 sec

9. *Click* **Finish** when you are done.

10. *Rename* the new satellite ✷ meo (which stands for **medium earth orbit**).

11. *Click* on the **3D Graphics Window** to make it active. *Click* the **Start** button on the toolbar 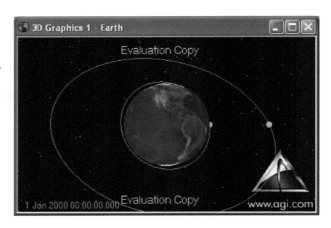 to animate the scenario. Watch the paths of the satellites you created.

12. When you finish, *click* the **Reset** button.

Satellite Graphics Properties
Now we will introduce some of the graphics properties available for a satellite and see how this affects the look of the satellite movement.

1. *Highlight* the ✷ **leo** satellite. *Right click* and *open* the Properties Browser. *Open* the **Attributes menus** under the **2D Graphics** folder.

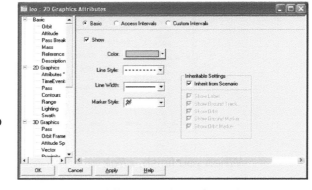

2. *Select* **Basic** and *change* the **Line Style** to `- - - - - - -` and the **Line Width** to `————`. *Click* **Apply** and notice the changes to the **2D Graphics Window**. *Reset* the **Line Style** to `————`.

3. Go to the **Pass Menu** and *enable* **Show Pass Labels** and **Show Path Labels**.

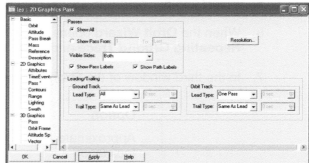

4. *Click* **Apply** to see the changes in the **2D Graphics Window**. Each groundtrack now shows the associated pass. Leave the leo graphics properties open.

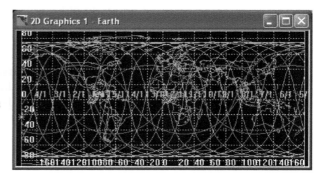

5. ***Click*** on the **3D Graphics Window** to make it active. ***Animate*** 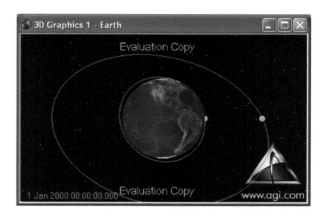 the scenario to see the effect of these changes. ***Press*** the **Reset** button when you have finished.

6. ***Change*** **Visible Sides** from **Both** to **Ascending**

    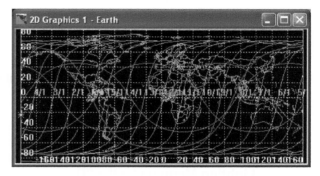 and ***click*** **Apply**. ***Click*** the **Start** button to see the changes in the **3D Graphics Window**. ***Press*** the **Reset** button when you are finished.

7. Now ***change*** **Visible Sides** back to **Both** and ***change*** the **Lead** and **Trail Types** under **Ground Track** from **All** to **Half**

    When you finish, ***click*** **Apply**.

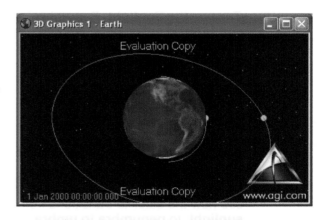

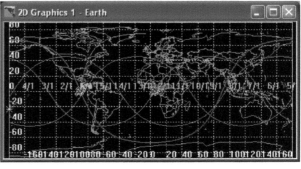

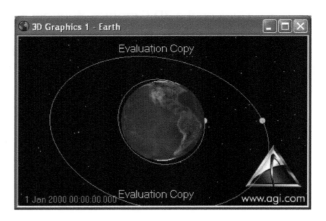

8. *Click* on the **3D Graphics Window** to make it active. *Animate* 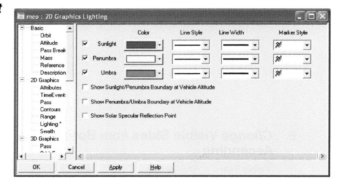 the scenario to see the effect of these changes. *Press* the **Reset** button when you have finished.

9. *Reset* the **Lead** and **Trail Types** under **Ground Tracks** to **All** and *disable* **Show Path and Pass Labels**. *Click* **OK**.

Satellite Lighting

You can display lighting conditions for individual satellites in the 3D **Graphics Window** by using the **Lighting** tool.

1. *Highlight* the ✖ meo satellite. *Right click* and *open* the 📋 Properties Browser . *Open* the **Lighting Menu** under the **2D Graphics** folder.

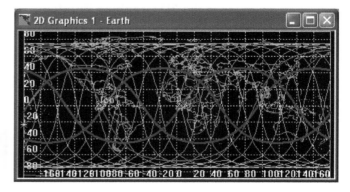

2. *Turn on* the **Sunlight**, **Penumbra**[1], and **Umbra**[2] options and *set* other options as follows:

Field	Color
Sunlight	Blue
Penumbra	White
Umbra	Red

3. *Apply* changes and *observe the* **2D Graphics Window**. *Magnify* 🔍 the ground track at a transition point to see how the satellite transitions from **sunlight**, to **penumbra to umbra**.

Zoom out to full view.

4. In the **Lighting** Menu, *enable* **Show**

---

[1] **Penumbra** *is the portion of a shadow that results from the source of illumination being only partially blocked.*
[2] **Umbra** *is the part of the shadow in which the light source is completely blocked.*
Wikipedia, The Free Encyclopedia. 9 Nov 2006, 02:15 UTC. Wikimedia Foundation, Inc. 10 Nov 2006

Sunlight/Penumbra Line at Vehicle Altitude and *click* **Apply**. *Reset* the **3D Graphics Window** and *animate* the scenario.

5. *Remove* all **Lighting** options and *click* [OK] to close the **Lighting** window.

Satellite Swath

The satellite swath displays field-of-view areas for a selected ground elevation angle or for a half angle relative to nadir or a surface distance. Swaths can be viewed with either the Edge Limits of the field-of-view shown, or the whole viewable area shaded. In this section, we will see both.

1. *Highlight* the 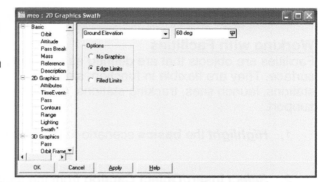 meo satellite. *Right click* and *open* the Properties Browser . *Open* the **Swath Menu** under the **2D Graphics** folder.

2. In the **Swath Menu**, *set* **Ground Elevation** to **60 deg**. *Select* **Edge Limits** and *click* **Apply** to view changes in the **2D Graphics window**.

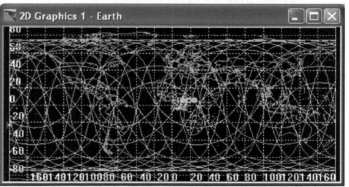

3. Now, *select* **Filled Limits** and *click* **Apply**. Notice the changes in the **2D Graphics window**. You can now see the entire field-of-view area represented by the shaded area.

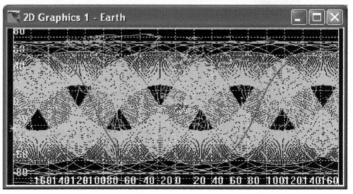

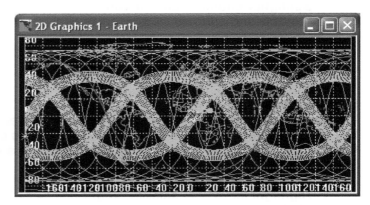

4. **Change** Ground Elevation Angle to **80 deg** and **Apply**.

5. Return the **Swath Menu** *to* `⊙ No Graphics` and *click* **OK**.

Working with Facilities

Facilities are objects that are defined as stationary locations on the Earth's surface. They are flexible in function, since they can be used to represent ground stations, launch sites, tracking stations, or other structures providing satellite support.

1. **Highlight** the **basics** scenario in the **Object Browser** window.

2. **Select** Insert/Object Catalog from the toolbar to **open** the **Object Catalog**. **Highlight** Facility and **click** Insert. If the **Orbit Wizard** appears, **click** Cancel to dismiss it. **Click** and **drag** across the name of the new facility so that the name may be edited. **Change** the name of the new facility to `🏠 socal`.

3. **Right click** `🏠 socal` and **open** the `📋 Properties Browser` for this facility. **Enter** the facility's exact position as follows:

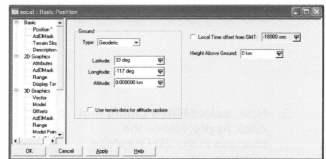

Option	Description
Type	Geodetic
Latitude	33 deg
Longitude	-117 deg
Altitude	0.000 km

4. **Click** `OK`. The new facility appears in the **2D Graphics window** at the specified location.

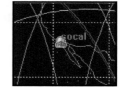

5. **Create** a second facility. **Rename** the facility **cape_canaveral**.

---

6. **Right click**  and **open** the 🗒 Properties Browser for this facility. **Enter** the facility's exact position as follows:

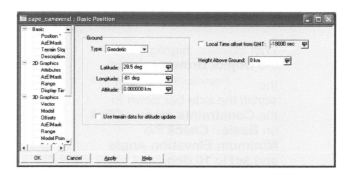

Option	Description
Type	Geodetic
Latitude	28.5 deg
Longitude	-81 deg
Altitude	0.000 km

7. **Click** ⬛ OK . The new facility appears in the **2D Graphics Window** at the specified location.

Determine Access & Apply Constraints

By determining accesses, you can find out when one object can see another object. In addition, you can impose constraints on accesses between objects to define what circumstances limit access.

1. **Highlight** the **socal** facility in the **Object Browser** window. From the Facility menu **select**  Access .

2. **Select** the 🛰 leo satellite from the **Associated Objects** and **click** the Access... button in the **Reports** section. This generates both the **Access** and the **Report**. Notice the ground-track changes in the **2D Graphics Window**. They are highlighted where the two objects have access.

3. Review the **Facility-socal-To-Satellite-leo - Access** report.

How many total seconds in duration is the total access period between the 🛰 leo satellite and the 🔒 socal facility? **Record** your answer on the **S1U5L1 Access-Constraint Worksheet**.

4. **Minimize** ⬛ the **Report** and **Access** windows for later use.

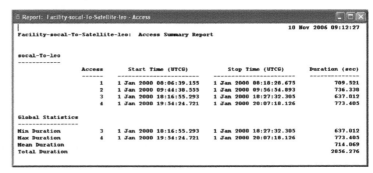

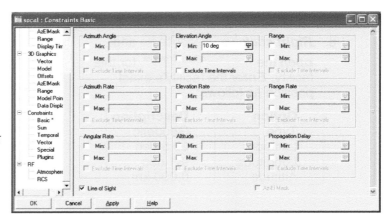

5. With 🔒 socal highlighted in the **Object Browser**, *open* the 📋 Properties Browser and *scroll* the side bar down to the **Constraints menu.** *Click* on **Basic.** *Check* the **Minimum Elevation Angle** and *set* to **10 deg**.

6. *Click* **Apply.** Notice the change in the **2D Graphics Window.** The access extent has been modified by the addition of constraints to the elevation angle at which the facility can access the satellite.

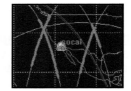

In this example, the 📡 leo satellite can now only be viewed from the 🔒 socal facility when it is 10 degrees above the horizon.

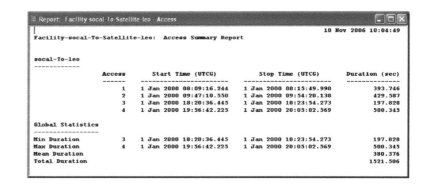

7. *Resize* 🗗 the report. From the **Main** menu, *select* **Refresh** from the **Report** menu in the **Report Window** and notice the change in access duration.

8. Review the revised report. How many total seconds in duration is the revised total access period between the 📡 leo satellite and the 🔒 socal facility? *Record* your answer on the **S1U1L16 Access-Constraint Worksheet.** *Close* ✖ the report.

9. In the **Constraints Menu,** *uncheck* the **Min Elevation** constraint and *click* **OK**.

10. *Resize* 🗗 the **Access** window and *click* the **Remove All** button. *Close* the **Access window** by *clicking* **Close.**

Add Sensors to the Scenario

Sensors can be used to represent such equipment as optical or radar sensors, receiving or transmitting antennas, or lasers. They can also be used to define another object's field of view. Although sensors are objects, they are subordinate to, or sub-objects of, the parent object to which they are attached.

1. ***Highlight*** the 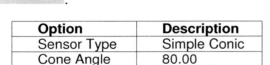 satellite in the **Object Browser** window, and ***insert*** a new sensor with the same method used to insert satellites and facilities. ***Rename*** the sensor track .

2. ***Right click*** track to open the Properties Browser window, and ***select*** the **Definition Menu** under **Basic**. ***Set*** the values listed below. ***Click*** OK .

Option	Description
Sensor Type	Simple Conic
Cone Angle	80.00

3. In the **Pointing Menu** for the **track** sensor, ***set*** the following options:

Option	Description
Pointing Type	Targeted
Boresight Type	Tracking

4. ***Highlight*** the socal facility from the **Available Targets** list and ***click*** on the **right-facing blue arrow** to add it to the **Assigned Target** list.

5. ***Repeat*** this process and ***move*** the cape_canaveral facility to the **Assigned Target** list. ***Click*** OK .

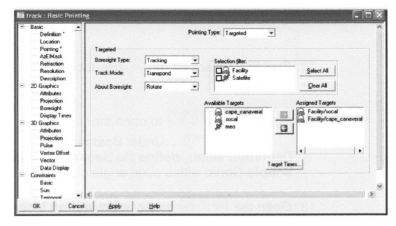

---

6. *Create* another sensor for the 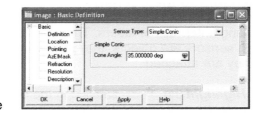 satellite, and *rename* it ■ image.

7. *Right click* ■ image to open the ▤ Properties Browser . Under **Basic** in the **Definition Menu**, *define* the **Sensor Type** as **Simple Conic** with a cone angle of **35 deg**.

8. *Scroll down* to the **2D Graphics** menu and *open* the **Attributes** menu for the sensor **image**. *Change* the **color** to be different from that of the previously created **track** sensor and *click*

    OK . This will help when viewing the sensor in the map window.

9. *Animate* the scenario in the **2D** and **3D Graphics Windows** by clicking on the **Start** button ▷ on the main toolbar. Notice that when the satellite gets within range of either facility, the track sensor pattern appears. The pattern of the image sensor, which is non-tracking, is always displayed.

    *Note: You can change the perspective of the 3D tool window by clicking on the **View Position** and **Direction** button and clicking on the satellite you wish to view from, as well as the point you wish to view.*

10. *Create* a sensor for the ☎ socal facility and *rename* it ■ uplink.

11. *Right click* ■ uplink to open the ▤ Properties Browser . Under **Basic** in the **Definition Menu**, *define* the **Sensor Type** as **Simple Conic** with a cone angle of **80 deg**.

12. *Open* the ■ uplink sensor's **Projection Menu** under **2D Graphics** in the **Properties Browser** and *scroll down* in the window to *set* the following values:

Option	Description
Persistence	0.00 sec
Minimum Altitude	0 km
Maximum Altitude	600 km
Step Count	1

13. ***Click*** OK . This sets the altitude of the sensor's projection to the height of the 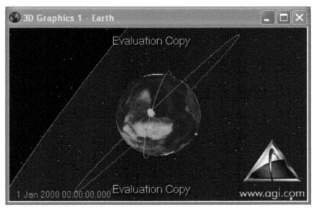 satellite.

14. ***Reset*** the **3D Graphics Window** and ***animate*** the scenario to observe the sensor displays over time. ***Reset*** the scenario when finished.

15. To save your training session for later review, ***click*** on the basics scenario in the **Browser** window. ***Click*** on the **File** menu and ***click*** **Save**. '

16. ***Select*** **Exit** from the **File** menu to exit the **STK** software.

Lesson 2:
STK and the GPS Satellite Constellation

One of the most widely used satellite and aerospace technologies with which the public interacts is the GPS System. Previous lessons referred to the GPS Constellation, which must be monitored to ensure everything is in working condition. This system is constantly evolving with older satellites being deactivated and used as "backups" as new technologies are inserted in to orbit. Planning is essential, and STK is one tool at engineers and technicians disposal to observer and map the expanding system.

The detailed model will contain specifications about the orbit, the name, owner, date inserted into orbit, mass, launch site, etc. The more details included in the analysis the better. Much like metadata for GIS, it is important to document all of the information available to ensure accuracy.

Looking at the GPS Constellation and individual GPS satellites also offer some new opportunities to learn in the STK environment. Specific satellites rely on specific inputs to accomplish their mission. With GPS satellites, they should be always directed towards the ground and like many satellites also need to maximize sunlight exposure. .

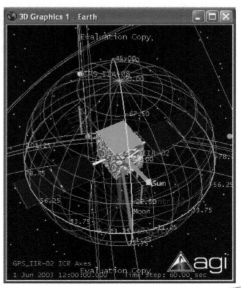

STK allows users to add **attitude spheres** to visualize the rotation of the satellite. Each satellite can rotate on 3 axes. Think of yourself diving off a diving board into a pool. You could do a front flip. You could spin. Or you could flip to the side, similar to a cartwheel. With practice you could move in all three of these directions, this would be spinning on all 3 axes. In space, with limited gravity, this is easily accomplished with little momentum. Satellites will rotate similarly, but we need to verify that this rotation is controlled so that the satellite is pointed at the Earth. Attitude spheres allow users to visualize that rotation with respect to the Earth.

In addition to knowing the attitude with respect to the Earth, we need to know where a few other points are. STK includes vectors to contantly show the direction of certain objects. In this lesson, we will focus on **nadir** or the position directly below the satellite. This is important to ensure that the satellite is focused towards the Earth. Vectors are represented by arrows. Other important vectors are the sun and velocity, speed in the direction of it's path, of the satellite. The velocity is necessary to know how when the satellite passes over specific ground points where receivers are located. The **sun** vector illustrates how the satellite's rotation is directly related to the sun's rotation. The satellite relies on the sun for its power.

One the signal has left the satellite it must interact with facilities on the ground. In GPS there are several key facilities. In this lesson, we will insert GPS monitoring stations into our GPS constellation scenario. The GPS constellation is constantly monitored from five locations around the world. These include Colorado Springs, Kwajalein, Diego Garcia, Ascension Island, and Hawaii. These stations monitor the satellites within range to make sure they are functioning properly. Colorado Springs(Schreiver Air Force Base) serves as the Master Control Station, the ONLY installation that can communicate with the satellites.

Using STK can show users how the GPS Constellation will be arranged at any epoch or acroos any time period. Understanding and visualizing vectors and attitude spheres, can ensure that the GPS constellation is functioning properly and to its fullest capabilities. With these tools, you are using industry standard aerospace software used by the military to accomplish these same tasks.

With GIS, users work toward representing the Earth two dimensionally to simplify location, direction, and measurements. When you map objects above the earth's surface that have varying altitudes and travel at high speeds, you must looks at the world in three dimensions. How exactly do the altitudes vary? There are a few basic ways that a satellite can travel around the Earth. In a **Low Earth Orbit (LEO)**, a satellite with a relatively short period (or orbits the Earth in 90 minutes) orbits the Earth at a height of less than 2,000 km above Earth's surface. A **Medium Earth Orbit** has an orbit up to 10,000 km.

Some satellites travel in circles, maintaining a close to constant altitude while traveling around the Earth. Some satellites very and travel in an elliptical orbit, meaning that they may pass close to the Earth and then travel to higher altitudes. In the image above we see two satellite paths that move around the Earth, but vary in their altitudes. The maximum point of distance from the Earth is called the **apogee**, while the minimum point is the **perigee.** The third type of orbit is the **Highly Elliptical Orbit**, which may travel up to a 50,000 km apogee and have perigee as close as 500 km.

In the previously described orbits, satellites are traveling faster than the Earth so to speak. In some situations a satellite may need to view the same portion of Earth for an extended period of time. In a **Geostationary orbit**, a satellite may orbit over a fixed position over the Earth. This can be useful in observing weather and holding extended periods of communication over one area.

The above orbits are in part based on the function of a satellite. How much does the satellite

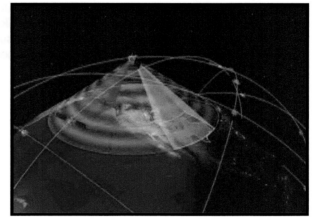

need to view? The **swath** is the area of coverage for a satellite's sensor. With a low orbit the satellite has a smaller swath. In higher orbits, the sensor has a larger swath...similar to the view of the Earth you have in an airplane versus on top of a small building. High Resolution imagery satellites usually have smaller swaths and are closer to the Earth for better quality imagery. By contrast, GPS satellites affect large groups of people simultaneously and have travel in higher orbits for a larger swath for maximum coverage. The same is true for telecommunication satellites.

What other considerations are there when traveling above and over the Earth? In the medium earth orbit definition, the satellite had a "short period" of 90 minutes or 90 minutes per revolution. In this case the satellite will actually pass out of the sun's range into "night" or a shadow of the Earth. When the satellite is in a shadow and completely out of sight of the sun, the satellite is in **umbra.** When it is in shadow and partially in view of the sun it is said to be in **penumbra**. The satellite's capabilities may change when in umbra or penumbra. For instance, a remote sensing satellite that takes natural color imagery is not useful without the sunlight illuminating the Earth's surface.

Another factor in Satellite modeling is time. TIme is a large factor in modeling and planning satellite paths. There are many satellites orbiting and many of these satellites need to make observations at a certain time. While working in STK, the **time period** is the total duration of the analysis. The **epoch** is the fixed instance that an event will occur that serves as a reference for all other times in the scenario. For instance, in a scenario moving a satellite to make an observation, the epoch may be the instance that an image is taken. All planning for the satellite must include activities that happen before and after that.

Before looking at the software, users need to know one last essential. Above concepts focused on satellite interaction with the Earth. Often satellites are used to communicate with facilities and receivers on the ground. These ground based objects have **constraints** and factors that limit **access**. A constraint is something that can limit the access that a facility or target has to a satellite sensor. That may be a tree or a building or anything else that could potentially obscure contact with a satellite. These constraints limit access or the time that one object can see or access another object.

When working with STK or in the satellite industry, knowledge of software is crucial, as is understanding the importance of planning the scenarios and processes that take place. STK like ArcGIS has the goal of communicating. One of the major requirements in communicating is

---

using industry specific terminology. The above terminology can help you adjust to industry terminology and ideas.

When applying this knowledge in STK, users have four basic objects to consider. Satellites. Facilities. Targets. Sensors. **Satellites** are vehicles that are placed in Earth's orbit to view and collect physical and human data about the Earth. **Facilities** are objects that are defined as stationary locations on the Earth's surface. They may represent ground stations, launch sites, tracking stations, or other structures providing satellite support. **Targets** are locations that are identified for observation by the satellite sensor like the coordinates that may be the location of an area to be imaged. **Sensors** are equipment such as optical or radar sensors, receiving or transmitting antennas, or lasers; sensors are subordinate to (sub-objects of) the object to which they are attached.

Lesson 2: STK GPS Satellite Constellation

Satellite Tool Kit (STK) is a multifaceted tool designed for aerospace professionals. In this lesson, you will use STK to explore the **GPS satellite constellation**. You will learn about the individual satellites in the constellation, and set and edit their properties.

1. To *launch* the **Satellite Tool Kit**, *click*

    on the **STK 9 Icon** .

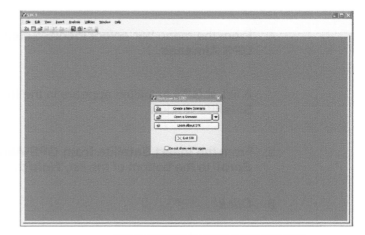

2. *Select* Open a Scenario
    from the STK menubar.

Select **File>Open** from the toolbar. When the dialog box opens, *select* **C:\STARS\IntroGISRSConcepts\GPS_Lesson.sc.**

It may take a moment to load. When it loads, the 2D and 3D graphics windows appear. *Resize* these windows so they are both visible.

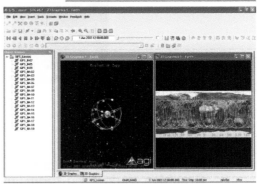

3. *Select* **File>Save As** from the menu bar and *save* the file in your student folder as **S1U6L2_XX (where XX is your initials).**

   This scenario that is loaded is setup with the active satellites – the **Block IIR** and **Block IIA** model satellites. We will now populate the most recently developed **Block IIR** satellites.

4. From the menu bar *select* **Insert>Satellite from GPS Almanac...**

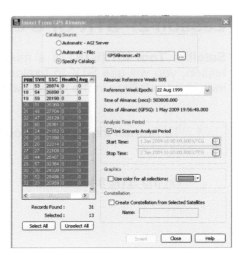

   A list of GPS satellites appears in the **Insert Satellite** window.

5. From the **Insert Satellite from GPS Almanac Window** window, *Select* **20** under PRN. *Scroll* to the bottom of the list. *Hold* the shift key and *click* **32.**

6. *Click*  .

7. *Click*  .

   Notice that the **Block IIR Satellites** now appear in the **Object Browser** and **3D Graphics** window along with the pre-loaded **Block IIR and IIA Satellites**.

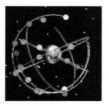 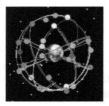

The GPS Constellation

1. *Right click, hold and drag* the cursor in the **3D Graphics** window and *zoom* until the view includes the whole GPS constellation.

2. *Click* the **Start** button  to animate the scenario. Watch the paths of the satellites you added to the scenario.

3. When you finish, *click* the **Reset** button ⟨◄◄ ◄ ◄ Ⅱ ► ►► ▼ ▲⟩.

The Constellation as a whole is comprised of six orbit paths that include at least four satellites a piece. This may vary depending on new satellite launches. Notice how the orbit path seems to shift. The orbit does not actually change; the earth rotates around the sun, and hence it's relative axis angle shifts, but since the view is focused on the earth, the orbit appears to shift accordingly.

Individual GPS Satellites

Setting Ground Track Properties:

1. *Right click* ✳ `gps-13_svn43` in the **Object Browser** and *open* the 🗒 Properties Browser .

2. *Click* anywhere on the **3D graphics** window to make it active.

3. To look more closely at that specific GPS satellite, *click* the **View From/To button** ⟨👁⟩
   on the **3D Graphics toolbar**. ⟨👁⟩ ⟨👁⟩ ⟨👁⟩ ⟨👁⟩

4. *Scroll down* to *select* ✳ `GPS_IIR-02` from the **View From** menu. The **View To** defaults to the same satellite selection.

5. **Click** ⟨ OK ⟩ .

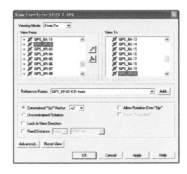

The **3D Graphics** window zooms to the ✳ `GPS_IIR-02` satellite in the GPS constellation.

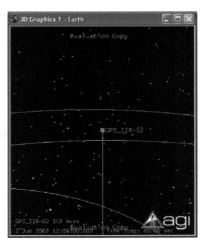

6. To rotate the view to position the satellite over the earth, *left click, hold and drag* the cursor *down* in the **3D Graphics** window.

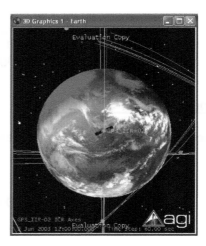

7. To get a closer look at the satellite, *right click, hold and drag* the cursor down to zoom in to the satellite.

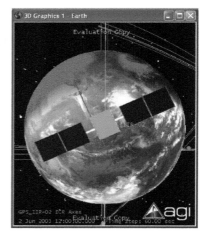

8. To rotate the view of the satellite, *left click, hold and drag* the cursor *down* to change the perspective of the view.

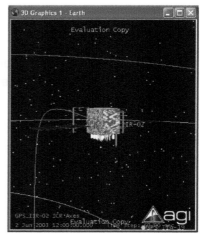

9. Return the view to show the zoomed in satellite hovering over the Earth.

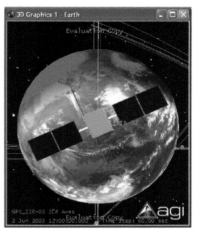

10. **Right click**  in the **Object Browser** and **open** the Properties Browser.

11. **Select** Pass under **3D Graphics**.

12. Under **Ground Track**, **select Full** as   Lead Type.

13. **Click** Apply **and** OK.

Notice that the **ground track** for **GPS 13** has been placed on the Earth as a circular path. This is the path that the satellite continually follows, but the satellite will not have a distinct swath in the same way as other satellites. The radio signal from the satellite travels in all directions like sound. Receiving the signal depends on the power of the GPS receiver.

16. **Click** the **Start** button ▐◀◀◀❙▷▷▷❤▲ to animate the scenario. Watch as the satellite follows it's ground track along the Earth's surface.

17. When you finish, **click** the **Reset** button ▐◀◀❙❙▷▷❤▲.

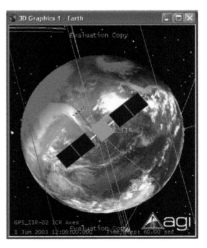

18. **Click** 🖫 from the toolbar to save your scenario.

Displaying Vectors

STK can display vectors that indicate the direction of **sun**, **nadir** *(the point on the Earth directly below the satellite)*, and **velocity** of a satellite. These indicators are helpful for managing the course and performance of a satellite.

1. ***Right click***  in the **Object Browser** and ***open*** the 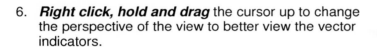 Properties Browser.

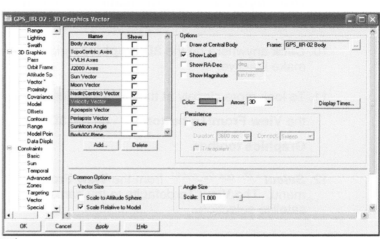

2. ***Select*** **Vector** under **3D Graphics**.

3. ***Check*** the boxes beside **Sun Vector**, **Nadir (Centric) Vector**, and **Velocity Vector**.

4. ***Select*** a different color for each vector.

5. **Click** Apply and OK.

6. ***Right click, hold and drag*** the cursor up to change the perspective of the view to better view the vector indicators.

   Notice that arrows are now extending from the satellite outward in three different directions. One points directly below towards the nadir position on the Earth. Nadir is important to GPS, since the signal needs to be focused directly at the Earth to gain the maximum signal strength.

   The other arrow shows the satellite's velocity, or speed in the direction of it's path. The velocity is necessary to know how when the satellite passes over specific ground points where receivers are located.

   The sun vector illustrates how the satellite's rotation is directly related to the sun's rotation.

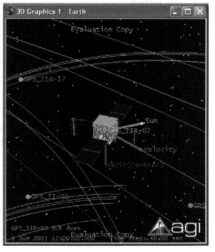

7. ***Click*** the **Start** button [buttons] to animate the scenario. Watch as the satellite rotates along it's path. Experiment with different perspective views of the satellite to see how it rotates and travels along it's ground track.

8. When you finish, ***click*** the **Reset** button [buttons].

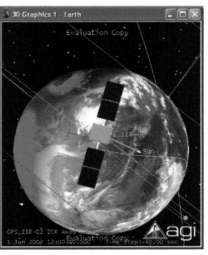

9. ***Use Steps 1-4*** to add **sun**, **nadir**, and **velocity** vectors for another satellite, ⚡ gps-02_svn61 .

10. ***Click*** anywhere on the **3D graphics** window to make it active.

11. To look more closely at this GPS satellite, ***click*** the **View From/To button** on the **3D Graphics toolbar**.

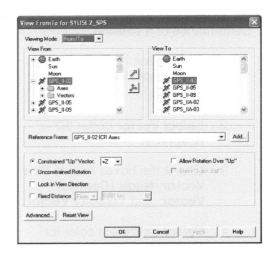

12. ***Select*** ⚡ gps-02_svn61 from the **View From** menu. The **View To** defaults to the same satellite selection.

13. ***Click*** OK .

14. ***Right*** and ***left click*** to zoom in and rotate the view perspective.

15. ***Click*** the **Start** button to animate this scenario. Watch as the satellite rotates to position it's solar panels to maintain maximum exposure to sunlight.

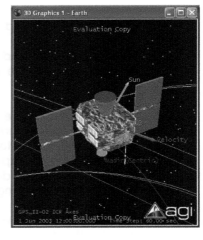

14. When you finish, ***click*** the **Reset** button .

15. ***Click*** 🖫 from the toolbar to save your scenario.

Generating Attitude Spheres

STK can generate and display the **attitude spheres** of each satellite. The **attitude** of a satellite is measured by "the angle the satellite makes with the object it is orbiting, usually the Earth. Attitude determines the direction a satellite's instruments face. The attitude of a satellite must be constantly maintained; this is known as attitude control."[1]

1. **Right click** gps-13_svn43 in the **Object Browser** and **open** the Properties Browser.

2. **Select Attitude Sphere** under **3D Graphics**.

3. **Check** the box beside **Show**.

4. **Click** Apply and OK.

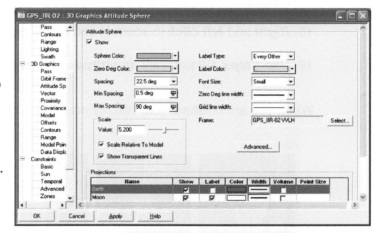

5. To zoom to gps-13_svn43, **click** anywhere on the **3D Graphics** window to make it active.

6. **Click** the **View From/To button** on the **3D Graphics** toolbar.

7. **Select** gps-13_svn43 from the **View From** menu. The **View To** defaults to the same satellite selection.

8. **Click** OK.

9. **Right** and **left click** to zoom in and rotate the view perspective.

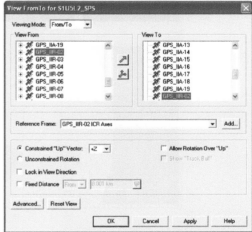

A wire sphere is added around the satellite to show the direction and the movements of the satellite with respect to it's angle to the Earth.

---

[1] *http://satellites.spacesim.org/english/glossary/ac.html*

---

16. To open a new Attiude window, *click* **the New 3D Attitude Graphics Window button** from the **STK Toolbar** .

17. *Resize* the window and position it to display in the upper left corner of the window.

18. *Right* and *left click* to zoom in and rotate the view perspective.

19. *Use Steps 1-9* to create an **attitude sphere** gps-02_svn61 .

When you are finished, your display may look similar to the display below:

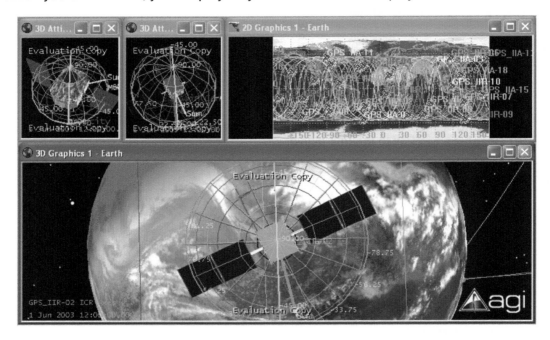

20. *Click* from the toolbar to save your scenario..

Inserting the GPS Ground Monitoring Stations

In this lesson, we will insert GPS monitoring stations into our GPS constellation scenario. The GPS constellation is constantly monitored from five locations around the world. These include Colorado Springs, Kwajalein, Diego Garcia, Ascension Island, and Hawaii. These stations monitor the satellites within range to make sure they are functioning properly. Colorado Springs(Schreiver Air Force Base) serves as the Master Control Station, the ONLY installation that can communicate with the satellites, communicating updated ephemeris data.

1. From the menubar, *select* **Insert>**
   Facility From Database…

2. *Check* the box beside **Site Name.**

3. *Type* **Hawaii** in the **Site Name** text box.

4. *Click* **Search.**

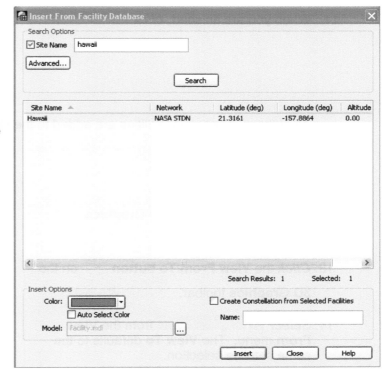

5. *Click* on the **Hawaii** station and *click* **Insert.**

6. *Type* **Diego Garcia** in the **Site Name Box of the Facility Database** window.

7. *Click* **Search.**

---

8. **Select** the facility in the **NASA STDN network**.

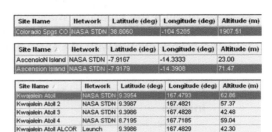

9. **Click**  OK .

10. Repeat this process for site names: **Colorado, Ascension Island,** and **Kwajalein**.

11. When you are finished, **click** Close .

All of these monitoring facilities are added to the **Object Browser**.

- Ascension_Island
- Colorado_Spgs_CO
- Diego_Garcia_Island
- Hawaii
- Kwajalein_Atoll

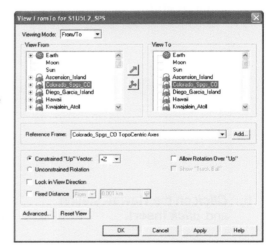

12. **Click** anywhere on the **3D Graphics** window to make it active.

13. **Click** the **View From/To button** on the **3D Graphics** toolbar.

14. **Select** Colorado_Spgs_CO from the **View From** menu. The **View To** defaults to the same satellite selection.

15. **Click** OK .

A ground view of this station appears in the **3D Graphics** window.

16. **Right** and **left click** to zoom in and rotate the view perspective to see this and the other stations from the global perspective.

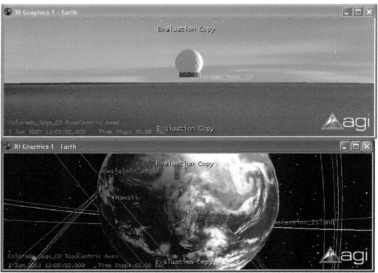

17. **Click** 🖫 from the toolbar to save your scenario and show your teacher your work.

18. If you wish to continue on to the next **STK** lesson, *select* **Close** from the **File** menu to close the current scenario. If you wish to exit the program at this time, *select* **Exit**.

GPS_IIR-02 ECF Position Velocity
Time (UTCG): 1 Jun 2003 12:00:00.000
x (km): .20116.245833
y (km): -10312.641612
z (km): -13931.090141
vx (km/sec): 1.766357
vy (km/sec): 0.094628
vz (km/sec): 2.496840

Lesson 3:
Reporting GPS Data with STK

When utilizing a tool such as STK, you are able to visualize many objects at one. When reporting your findings, users can communicate more effectively by using written data and reports in your presentation. STK allows users to report directly from within the software.

The first type of reporting available to users, we will call real-time reporting. STK allows users to display certain data about the satellite that is constantly updated. For this lesson we will focus on two pieces of data: EDF and RIC. The **ECF** *(Earth Centered Fixed)* **Position** shows the speed of the satellite measured as *velocity in kilometers per second*. This Earth-centered system locates objects in the solar system in three-dimensions along the Cartesian X, Y and Z axes. These satellites are traveling fast enough that in order to be accurate all three measures must considered when moving in space. The **RIC** shows a series of distance measurements, to include **radial, in-track, cross-track** and **range,** between designated satellites at a particular time.

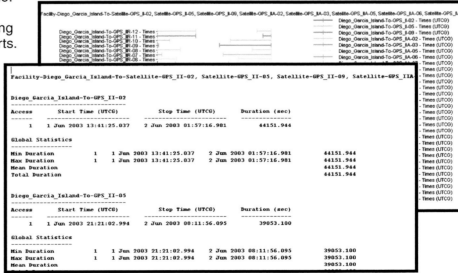

The second type of reporting available in is printed reports. Most notably access reports allow users to show access from a specific location over a period of time. The data report can appear in the form of a graph, which visually shows access. The report can also be written to include detailed times of access time or duration.

Spatial Technology And Remote Sensing
STARS Remote Sensing Education
SPACESTARS
Laboratory for GIS/Remote Sensing Education

Using both of these tools can enhance the effectiveness of presentations to interested stakeholders. Aerospace technology relies on accuracy. Reporting can provide detailed information about either a satellite with respect to its surroundings or about a facility with respect to other satellites. In this case, we will analyze access to the GPS constellation.

Lesson 3: Reporting GPS Data with STK Exercise

This exercise demonstrates the some of the powerful display and reporting features of the **STK** software. In this lesson you will:
- Create **Dynamic Displays** for **Velocity** and **RIC Data.**
- Create a report and graph for **access** from a **GPS ground station.**

1. To *launch* the **Satellite Tool Kit**, *click*

   on the **STK 9 Icon** .

2. *Select* 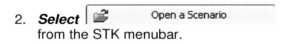 from the STK menubar.

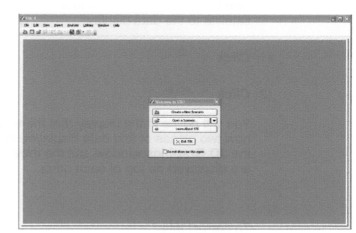

3. When the dialog box opens, navigate to your student folder, and select **S1U6L2_XX (where XX is your initials).**

   It may take a moment to load.

4. *Select* **File>Save As** from the menubar and *save* it in your student folder as **S1U2L3_XX (where XX is your initials).**

5. *Right click* 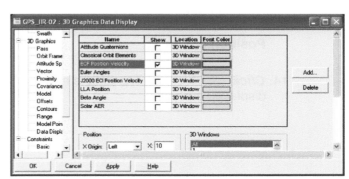 in the **Object Browser** and *open* the 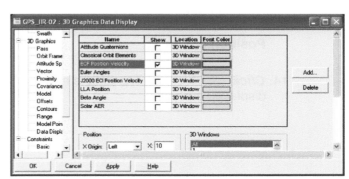 Properties Browser .

6. *Select* **Data Display** under **3D Graphics.**

7. *Check* the box to show **ECF Position Velocity.**
   The **ECF** *(Earth Centered Fixed)* **Position** shows the speed of the satellite measured as *velocity in kilometers per second*. This Earth-centered system locates objects in the solar system in three-dimensions along the Cartesian X, Y and Z axes. These satellites are traveling fast enough that in order to be accurate all three measures must considered when moving in space.

8. *Click*  .

---

9. ***Select* RIC** and ***click*** 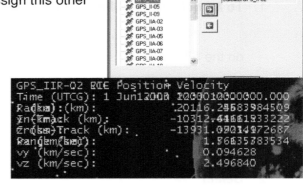 OK .

The **RIC** shows a series of distance measurements, to include **radial, in-track, cross-track** and **range,** between designated satellites at a particular time.

10. ***Select*** ✷ gps-02_svn61 and ***click*** ⬅ to assign this other satellite to the data display.

11. ***Click*** OK .

12. ***Click*** Apply .

You will notice in the display that there is now two pieces of text data displayed in the **3D Graphics** window, but the they are displayed on top of each other.

To correct this,

13. ***Click*** on **RIC** in the **3D Graphics Data Display** window and input **125** into the

Position box for **Y:**.

14. ***Click*** OK . The data is now displayed correctly.

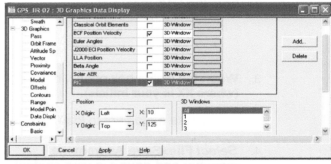

Viewing Access

1. *Right click*  Diego_Garcia_Island and *select* **Access Tools**.

2. *Select* ⚹ gps-01_svn49.

3. *Scroll* to the bottom and *while holding the Shift Key, click* ⚹ gps-32_svn23.

4. *Click* Access... under **Graphs.**

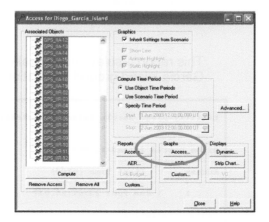

5. *Double click* on the window title bar to maximize the graph window;

▥ Graph: Access - Access   ⎕ ⧉ ✕

and, if necessary, adjust the **Object Browser** window to fully maximize the graph report window.

This report illustrates graphically the access time to each satellite from the Diego Garcia Island ground installation:

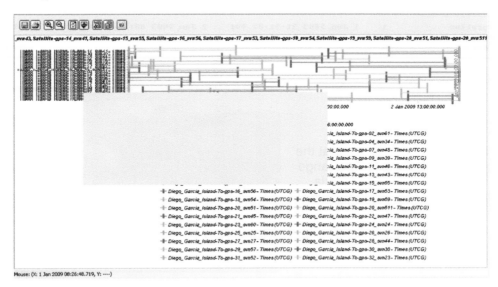

6. ***Click*** the **Access for Diego Garcia Island window** tab [Access for ...] at the bottom of the display.

7. This time ***click*** [Access...] under **Reports**.

> Reports
> Access...
> AER...
> Link Budget
> Custom...

This report displays numerically the exact access time to each satellite from the Diego Garcia Island ground installation:

```
Facility-Diego_Garcia_Island-To-Satellite-GPS_II-02, Satellite-GPS_II-05, Satellite-GPS_II-09, Satellite-GPS_IIA-

Diego_Garcia_Island-To-GPS_II-02
--------------------------------
Access       Start Time (UTCG)          Stop Time (UTCG)          Duration (sec)
------       -----------------          ----------------          --------------
    1    1 Jun 2003 13:41:25.037    2 Jun 2003 01:57:16.981         44151.944

Global Statistics
-----------------
Min Duration        1    1 Jun 2003 13:41:25.037    2 Jun 2003 01:57:16.981         44151.944
Max Duration        1    1 Jun 2003 13:41:25.037    2 Jun 2003 01:57:16.981         44151.944
Mean Duration                                                                       44151.944
Total Duration                                                                      44151.944


Diego_Garcia_Island-To-GPS_II-05
--------------------------------
Access       Start Time (UTCG)          Stop Time (UTCG)          Duration (sec)
------       -----------------          ----------------          --------------
    1    1 Jun 2003 21:21:02.994    2 Jun 2003 08:11:56.095         39053.100

Global Statistics
-----------------
Min Duration        1    1 Jun 2003 21:21:02.994    2 Jun 2003 08:11:56.095         39053.100
Max Duration        1    1 Jun 2003 21:21:02.994    2 Jun 2003 08:11:56.095         39053.100
Mean Duration                                                                       39053.100
```

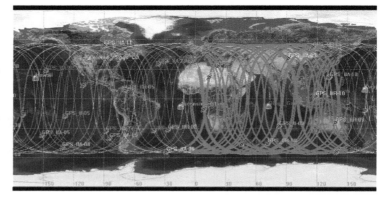

8. Click the [2D Graphics ...] tab at the bottom of the display. The range of access for the Diego Garcia Island facility is highlighted in red.

9. ***Click*** [💾] from the toolbar to save your scenario and show your teacher your work.

10. If you wish to continue on to the next **STK** lesson, ***select* Close** from the **File** menu to close the current scenario. If you wish to exit the program at this time, ***select* Exit**.

Lesson 4:
Simulating an Air Photo Mission

There are a variety of vehicles that have been used for remote sensing…all of these traveling above the Earth. STK allows you to model aircraft as well. Aircraft missions can be reconnaissance, travel, exploration, and, for our purposes, gathering aerial photos.

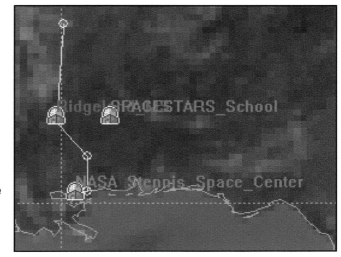

Since aircraft fly at much lower altitudes than satellites, air photos generally cover a smaller area than satellite imagery. Usually an order to task an air photo comes from a local source; for example, someone who is interested in developing a piece of property along a commercial corridor. Government agencies do task wider range aerial surveys for agricultural purposes, such as the NAIP Program, which offers an extensive library of high resolution aerial photography over much of the United States. The STK Toolkit can be used to task, manage and operate an air photo mission much in the same way that it can be used for a satellite mission.

Each situation has a vehicle traveling above the Earth. That vehicle must interact with facilities targets on the ground. The aircraft contains a sensor like a satellite that measures data, in this case imagery.

The differences in the two scenarios is how the aircraft plan is modeled. With the satellite scenario users specified orbital characteristics. Aircraft settings involve adding a flight plan for the aircraft by specifying location, speed, altitude, and time.

For this lesson, you will complete a scenario that models a flight from the gulf coast to Memphis, TN and then to your own school.

Lesson 4: Simulating an Air Photo Mission Air Photo Exercise

Since aircraft fly at much lower altitudes than satellites, air photos generally cover a smaller area than satellite imagery. Usually an order to task an air photo comes from a local source; for example, someone who is interested in developing a piece of property along a commercial corridor. Government agencies do task wider range aerial surveys for agricultural purposes, such as the NAIP Program, which offers an extensive library of high resolution aerial photography over much of the United States. The STK Toolkit can be used to task, manage and operate an air photo mission much in the same way that it can be used for a satellite mission.

View the Simulation

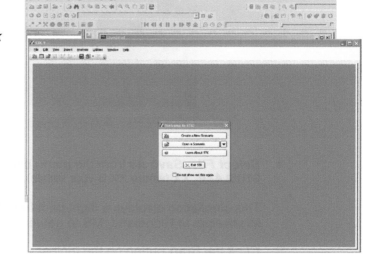

1. To *launch* the **Satellite Tool Kit**, *click*

    on the **STK 9 Icon** .

2. *Select* | Open a Scenario from the STK menubar.

    Select **File>Open** from the toolbar. When the dialog box opens, *select* **C:\STARS\IntroGISRSConcepts\Air_Photo.sc**.

.

3. When the image file loads, *click* the 2D Graphics ... tab to view the map display.

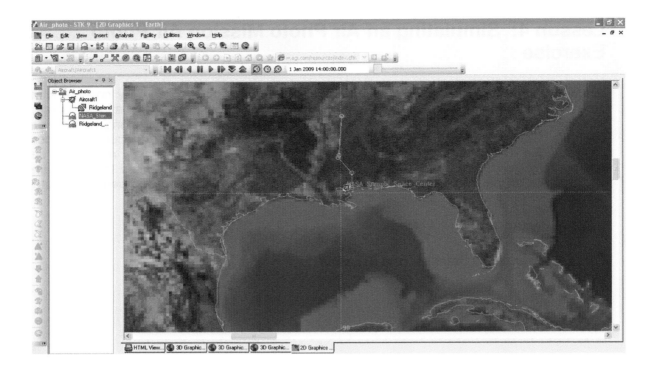

4. **Select File>Save As** from the menu bar and **save** in your student folder as **S1U6L4Air_XX** (where XX is your initials).

   This simulation displays a flight path from NASA's Stennis Space Center in south Mississippi to Ridgeland, MS to gather imagery data, and then on to Memphis, TN.

5. **Click** the **Start** button on the toolbar
   to animate the scenario. Watch the path of the aircraft tasked for this simulation in the 2D Graphic... and 3D Graphics ...windows.

   You will see data displays in the **3D Graphics** displays in yellow for the airplane: one for longitude and latitude, and one for velocity. The data for the Ridgeland sensor is displayed in red. When the range reaches zero, the aircraft is directly over the target.

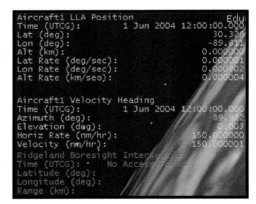

6. **Click** the **Reset** button after viewing the simulation.

7. If the animation runs too slow or too quickly, **click** the **Time Step** buttons , and repeat the animation.

Add Your Location as a Facility

Facilities are objects that are defined as stationary locations on the Earth's surface. They are flexible in function and can be used to represent ground stations, launch sites, tracking stations, or other structures providing satellite support. We will add your school to this scenario as a facility.

1. From the Insert menu, *click* **Object Catalog**. *Highlight* Facility and *click* Insert .

2. *Rename* the facility with your school name *(eg.* SPACESTARS_School *.)*

3. *Right click* **YourSchoolName** and *select* Properties Browser . *Enter* the your school's exact longitude and latitude coordinates that where recorded in earlier lessons as follows:

   (Note: You may need to convert the coordinates to decimal degrees, if not already done so.)

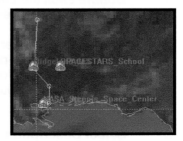

4. *Click* OK . Your school facility appears in the map window at the specified location.

Add Your Location to the Flight Path

Now that your school facility is added to the scenario, it can now be added to the flight path.

1. **Right click** 🛩 Aircraft1 and **select** 📑 Properties Browser .

   The route window appears listing the coordinates along the flight path.

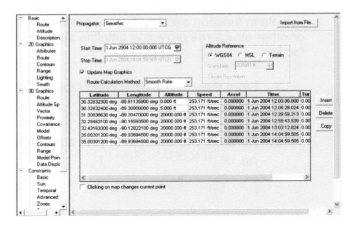

2. **Select** the last record in the list.

   | 35.00301200 deg | -89.93694500 deg | 20000.000 ft | 253.171 ft/sec | 0.000000 | 1 Jun 2004 14:04:59.505 | 0.0(|

   This is the last location, or point, on the flight path.

3. **Click** [Insert Point] to add a new point to the route list. An identical point is added to the list after the last record. This will be your school location on the flight path.

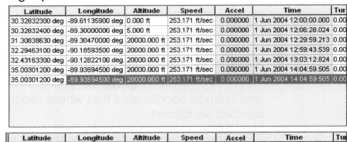

Latitude	Longitude	Altitude	Speed	Accel	Time	Tur
30.32832300 deg	-89.61135900 deg	0.000 ft	253.171 ft/sec	0.000000	1 Jun 2004 12:00:00.000	0.00
30.32832400 deg	-89.30000000 deg	5.000 ft	253.171 ft/sec	0.000000	1 Jun 2004 12:06:28.024	0.00
31.30838630 deg	-89.30470000 deg	20000.000 ft	253.171 ft/sec	0.000000	1 Jun 2004 12:29:59.213	0.00
32.29463100 deg	-90.18593500 deg	20000.000 ft	253.171 ft/sec	0.000000	1 Jun 2004 12:59:43.539	0.00
32.43163300 deg	-90.12822100 deg	20000.000 ft	253.171 ft/sec	0.000000	1 Jun 2004 13:03:12.824	0.00
35.00301200 deg	-89.93694500 deg	20000.000 ft	253.171 ft/sec	0.000000	1 Jun 2004 14:04:59.505	0.00
35.00301200 deg	-89.93694500 deg	20000.000 ft	253.171 ft/sec	0.000000	1 Jun 2004 14:04:59.505	0.00

4. **Double click** the **Latitude** field to enter the coordinates of your facility.

5. **Note** the **Time** entered in the **Time** field on a separate piece of paper. You will use this information later in the lesson when you animate the simulation.

Latitude	Longitude	Altitude	Speed	Accel	Time	Tu
30.32832300 deg	-89.61135900 deg	0.000 ft	253.171 ft/sec	0.000000	1 Jun 2004 12:00:00.000	0.0(
30.32832400 deg	-89.30000000 deg	5.000 ft	253.171 ft/sec	0.000000	1 Jun 2004 12:06:28.024	0.0(
31.30838630 deg	-89.30470000 deg	20000.000 ft	253.171 ft/sec	0.000000	1 Jun 2004 12:29:59.213	0.0(
32.29463100 deg	-90.18593500 deg	20000.000 ft	253.171 ft/sec	0.000000	1 Jun 2004 12:59:43.539	0.0(
32.43163300 deg	-90.12822100 deg	20000.000 ft	253.171 ft/sec	0.000000	1 Jun 2004 13:03:12.824	0.0(
35.00301200 deg	-89.93694500 deg	20000.000 ft	253.171 ft/sec	0.000000	1 Jun 2004 14:04:59.505	0.0(
32.44818900 deg	-88.67130000 deg	20000.000 ft	253.171 ft/sec	0.000000	1 Jun 2004 15:11:17.591	0.0(

6. **Select** Description from the **Basic** menu. **Enter** the name of your school in the **Short Description** field.

Short Description:	SPACESTARS School

7. **Click** [Apply] and [OK] .

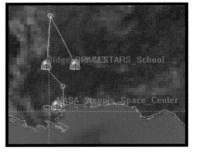

Now you can see your school site added to the flight path of the aircraft.

Add a Sensor to the Aircraft

Aircraft that fly to obtain aerial photography can be equipped with various types of imagery sensors. These can include panchromatic cameras that capture imagery in the visible bands of the electromagnetic spectrum; to infrared or thermal sensors that capture imagery from portions of the spectrum beyond the visible range. Each sensor's properties need to be set in STK to enable a true and accurate simulation based on the capabilities of the selected sensor.

1. ***Select*** Aircraft1 in the **Object Browser** window, and ***insert*** a new Sensor.

2. ***Rename*** the sensor **YourSchoolName_Data**.
   (eg. SPACESTARS_School_Data *)*.

3. ***Right click*** the new sensor and ***select*** Properties Browser.

4. Under **Basic** in the **Definition Menu** *set* the values as follows:

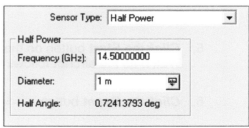

For this example, we assume that the plane is following a radio signal, similar to a beacon.

5. ***Click*** Apply.

6. ***Click*** **Pointing** under **Basic**.

7. ***Select*** **Targeted** as **Pointing Type**.

8. ***Select*** **Tracking** as **Boresight Type**.

9. ***Highlight*** your facility as an **Available Target** and ***click*** the add arrow to move it to the **Assigned Targets** list.

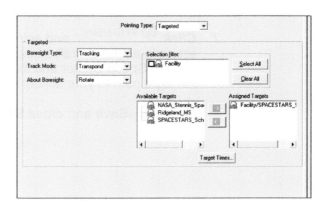

10. ***Click*** Apply and OK.

View the Simulation

1. From the **Object Browser** window, *right click* your scenario file *(eg.* 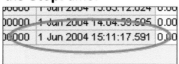 *)* and *select* 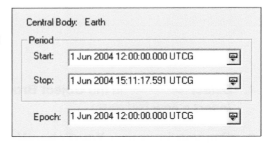 Properties Browser .

2. *Select* **Time Period** under the **Basic** menu and *set* the time that you recorded earlier as the **Stop:** time.

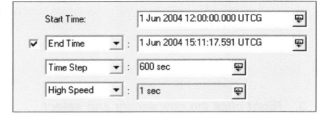

3. *Select* **Animation** under the **Basic** menu and set the **End Time:** as the time you wrote down.

4. *Click* Apply and OK .

5. *Click* the **Start** button on the toolbar to animate the **2D scenario**. Watch as the aircraft follows the path from tasked for this simulation.

6. *Click* the **Reset** button after viewing the simulation.

7. *Click* 3D Graphics ... and animate this scenario as the aircraft takes off from Stennis Space Center and flies along the flight path to your school location.

8. *Click* **File>Save** and *close* STK.

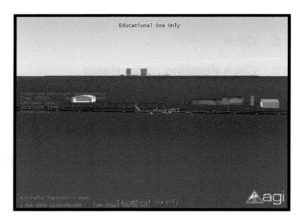

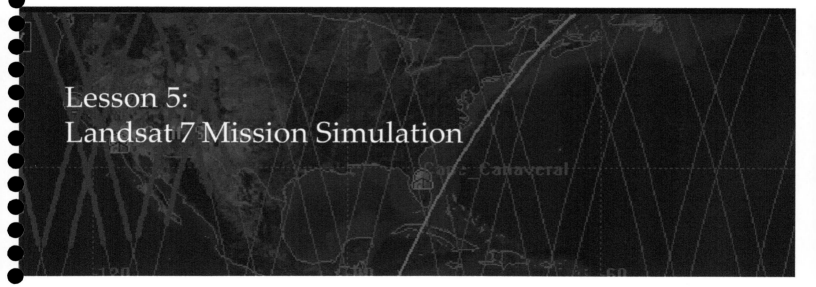

Lesson 5:
Landsat 7 Mission Simulation

NASA's Landsat program is the longest running image acquisition program in history. The Landsat data archives now contain almost 27 years of material. The data is collected at a 30 meter resolution in eight bands of the electromagnetic spectrum, to include visible, infrared and thermal bands. Landsat 7 repeats its ground trace every 16 days, and is separated from the earlier launched Landsat 5 path by eight days. In this lesson, you will use STK to simulate a Landsat 7 mission.

STK is a useful tool in modeling Landsat's mission. In addition, to Landsat seven there are thousands of objects orbiting the Earth, some of these objects could affect (or be affected by) the Landsat satellite. STK becomes a crucial tool in modeling the Landsat 7 sensor for its duration in space to ensure its safety as well as safety for other satellites.

In this lesson, users will combine the skills seen in the previous lesson to simulate Landsat as it gathers imagery.

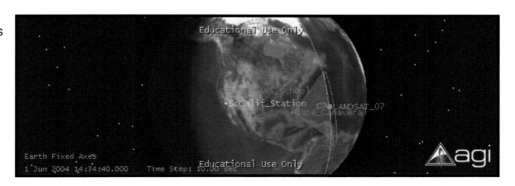

Lesson 5: Simulating a Satellite Mission Landsat Satellite Simulation Exercise

Set the Landsat 7 Scenario

1. To *launch* the **Satellite Tool Kit**, *click* on the **STK 9 Icon** .

   You will see the HTML viewer.

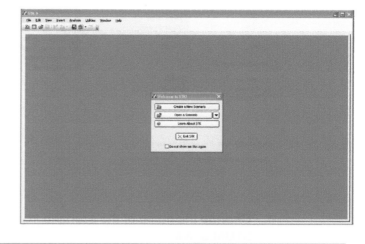

2. *Select* 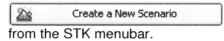 from the STK menubar.

3. **Name** the scenario **S1U6L5_XX** (where XX is your initials). Ensure that the Location is your student directory.

---

4. The **2D** and **3D Graphics Windows** open. Close the **Insert STK objects** window.

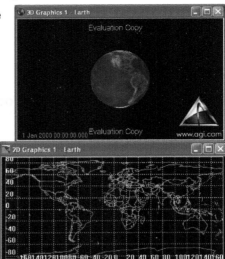

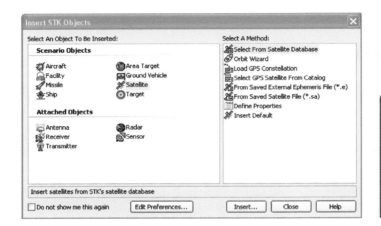

4. *Change* the name of the scenario from **S1U6L5_XX** to 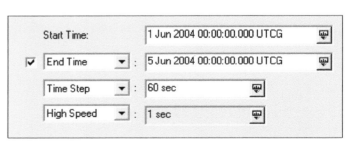 Landsat_Scenario.

5. *Right click* on Landsat_Scenario and *select* Properties Browser .

6. *Select* **Time Period** under **Basic** and *set* the following options:

    Note: Each scenario has a time period and an epoch. The time period defines the general time span for analysis. The epoch serves as a reference for all other times in the scenario.

Central Body: Earth

Period	
Start:	1 Jun 2004 00:00:00.000 UTCG
Stop:	5 Jun 2004 00:00:00.000 UTCG

Epoch: 1 Jun 2004 00:00:00.000 UTCG

7. *Select* **Animation** under **Basic** and *set* the following options:

Start Time:	1 Jun 2004 00:00:00.000 UTCG
✓ End Time ▾ :	5 Jun 2004 00:00:00.000 UTCG
Time Step ▾ :	60 sec
High Speed ▾ :	1 sec

8. *Select* **Units** under **Basic** and *confirm* the following options:

Dimension	CurrentUnit	UnitA
DistanceUnit	Kilometers (km)	km
TimeUnit	Seconds (sec)	sec
DateFormat	Gregorian UTC (UTCG)	UTCG
AngleUnit	Degrees (deg)	deg
MassUnit	Kilograms (kg)	kg

9. *Select* **Description** under **Basic** and, in the **Short Description** field, *type* **Landsat Imagery**.

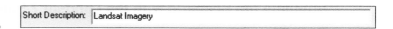

10. *Click* [OK] .

11. *Click* anywhere on the  window to ensure that it is selected.

12. *Click* the **Properties** button 🔲 on the Toolbar.

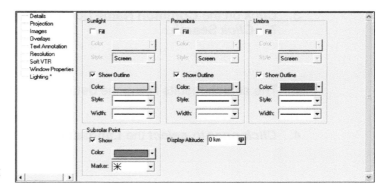

13. *Click* **Lighting**, and *enable* ☑ Show Outline **for Sunlight, Penumbra, Umbra, and** ☑ Show **the Subsolar Point** options. Select a different color for each property.

14. *Click* [Apply] and [OK] .

Populate the Scenario with the Landsat 7 Satellite

Define the Landsat 7 Satellite:

1. *Highlight* 🗺 Landsat_Scenario in the **Browser** window.

2. *Select* Insert > Satellite From Database... .

3. *Click* on the **Common Name** box and *type* in **Landsat 7**. *Click* **Search**.

    The search results listing the details of the **Landsat 7** satellite appears.

4. *Click once* to *select* the **Landsat 07** satellite. *Click* **Insert**.

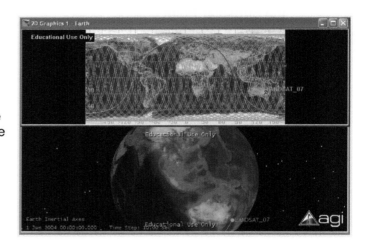

5. *Click* Close to *close* the Satellite Database window.

6. *Click* the **Start** button on the toolbar ▐◀ ◀◀ ◀ ❙❙ ▶ ❙▶ ⯆ ⯅ to animate the scenario. Watch the path of the Landsat satellite in the 🖼 2D Graphic... and 🌐 3D Graphics .. windows.

7. *Click* the **Reset** button ▐◀ ❙❙ ◀ ❙❙ ▶ ❙▶ ⯆ ⯅ after viewing the simulation.

8. If the animation runs too slow or too quickly, *click* the **Time Step** buttons ▐◀ ◀◀ ◀ ❙❙ ▶ ❙▶ ⯆ ⯅ , and repeat the animation.

Add Ground Station Facilities:
Facilities are objects that are stationary locations on the Earth's surface. They are flexible in function, and can be used to represent ground stations, launch sites, tracking stations, or other structures providing satellite support. We will add two ground stations to this Landsat scenario.

1. *Highlight* 🗺 Landsat_Scenario in the **Browser** window.

2. *Select* Insert > New... .

Introduction to GIS and RS Concepts, version 7

3. **Select** 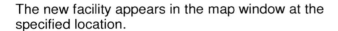 and **click** [Insert].

4. **Rename** the facility 🏛 SoCalif_Station.

5. **Right click** 🏛 SoCalif_Station and **select**
   📋 Properties Browser .

6. **Click** the **Position** menu.

7. **Enter** the facility's exact latitude and longitude position as follows:

8. **Click** [OK].

The new facility appears in the map window at the specified location.

9. **Follow** the same steps to create another ground facility named **Cape_Canaveral**. Enter the facility's exact position as follows:

   The new facility appears in the map window at the specified location.

10. **Zoom in** to view both ground stations.

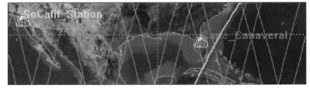

Add your school site as a target:
To monitor when the satellite acquires imagery data of your school site, the site is to be added to the scenario as a target.

1. *Highlight*  in the **Browser** window.

2. *Select* Insert > New... .

3. *Select* ⊙ Target and *click* Insert .

4. *Rename* the target with **your school name** *(eg.* ⊙ SPACESTARS_School *).*

5. *Right click* your **school target** ⊙ and *select*
    Properties Browser .

6. *Click* the **Position** menu.

7. *Enter* the target's latitude and longitude position as recorded in an earlier lesson:

8. *Click* the **Description** menu, and *enter* the name of your school.

Short Description: SPACESTARS_School

9. *Click* Apply and OK .

    The new target appears in the map window at the location of your school site.

Add Sensors:

Sensors are mounted on satellites to collect the image data. There are various types of sensors, including optical and radar sensors, receiving and transmitting antennas, and lasers.

1. **Highlight** 🛰 LANDSAT_07 in the **Browser** window.

2. **Select** Insert > New... .

3. **Select** 📡 Sensor and **click** Insert .

4. **Rename** 📡 Sensor1 to 📡 Data1 .

5. **Right click** 📡 Data1 and **select** 📋 Properties Browser .

6. **Click** the **Definition** menu.

7. **Set** the values as follows:

8. **Click** Apply and OK .

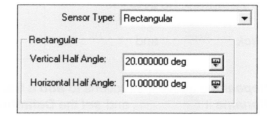

This sensor is added to the Landsat satellite.

To add a **tracking sensor** for the two ground monitoring facilities:

1. **Highlight** the 📡 Cape_Canaveral facility

2. **Select** Insert > New... .

3. **Select** 📡 Sensor and **click** Insert .

4. **Rename** 📡 Sensor2 to 📡 Tracker1 .

5. **Right click** 📡 Tracker1 and **select** 📋 Properties Browser .

6. **Click** the **Definition** menu.

7. **Set** the values as follows:

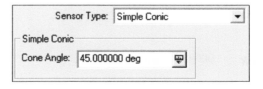

8. ***Click*** the **Pointing** menu.

9. ***Select*** **Targeted** as the **Pointing Type**.

10. ***Single click*** on the  LANDSAT_07 satellite to ***highlight*** and ***click*** to move the satellite into the **Assigned Targets** list.

11. ***Click*** Apply and OK .

12. ***Repeat*** the above instructions to ***add*** a **tracking sensor** for the SoCalif_Station facility ***Rename*** it Tracker2, and ***set*** the **Definition** and **Properties** values the same as Tracker1.

13. ***Use*** the left mouse button to ***center*** the **3D Graphics** window over **your school**.

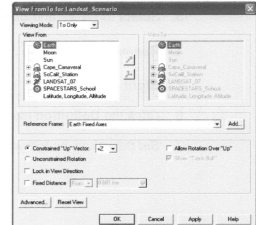

14. ***Click*** the **Set View Position and Direction** button .

15. ***Choose*** Viewing Mode: To Only and .

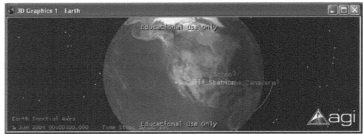

16. ***Click*** Apply and OK .

17. ***Click*** the **Start** button on the toolbar 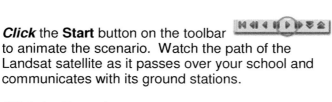 to animate the scenario. Watch the path of the Landsat satellite as it passes over your school and communicates with its ground stations.

18. ***Click*** the **Reset** button after viewing the simulation.

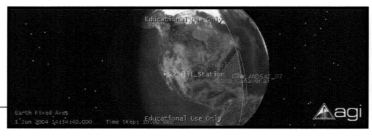

19. If the animation runs too slow or too quickly, *click* the **Time Step** buttons

, and repeat the animation.

Determine Access & Apply Constraints:

By determining accesses properties, you can find out when the **Landsat 7** satellite can see your school. Let's see how often in this scenario you can get data for your school.

1. *Right click* your **school target** ⊙ in the **Object Browser**.

2. From the Target Tools menu *select* **Access Tools**.

3. *Select* the ✳ LANDSAT_07 from the **Associated Objects** and *click* the Access... in the **Reports** section.

This will generate both the **Access** and the **Report**.

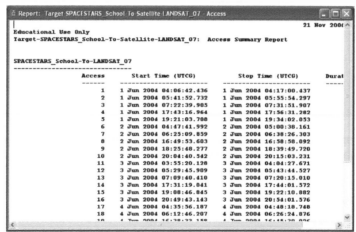

4. How many total seconds in duration is the total access period between the satellite and the facility? *Record* your answer on the **S1U5L5 Landsat 7 Access-Constraint Worksheet**.

5. *Minimize* the **Access Report** and **Access window** for later use.

Notice the ground track changes in the 2D Graphics window are highlighted where the satellite and the school are within access range.

---

6. **Right click** your **school target** 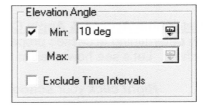 in the **Object Browser** and **select**
   Properties Browser .

7. **Scroll down** and **click** Basic in the **Constraints** menu.

8. **Enable** the **Minimum Elevation Angle** and **set** to **10 deg**.

9. **Click** Apply and OK .

   Notice the ground track changes in the **2D Graphics** window. The extent of access where the satellite and the school are within range is now limited by the constraint of access modified to only when the satellite is 10 degrees above the horizon .

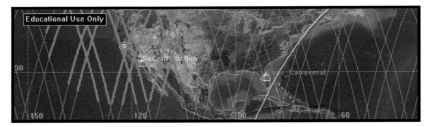

10. **Resize** the report. In the **Report** window, **select** **R**efresh from the **Report** menu and notice the change in access duration.

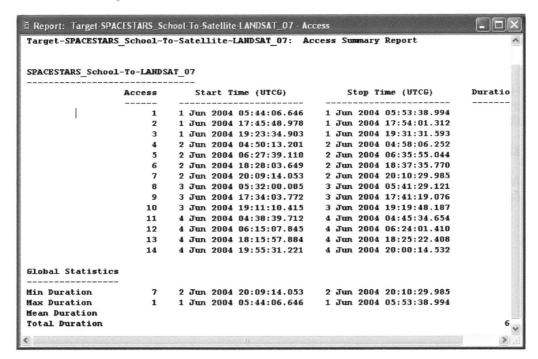

Now, how many total seconds in duration is the total access period between the satellite and the school site?

11. To save your scenario, *click* 🖼 Landsat_Scenario in the **Browser** window. *Click* the **File > Save as,** enter **S1U6L5SAT_XX.sc** (where XX is your initials) in your student folder.

12. *Select* **Exit** from the **File** menu to *exit* the **STK** software.

Appendix A

Washington DC, Reference Map
Unit 4, Lesson 4

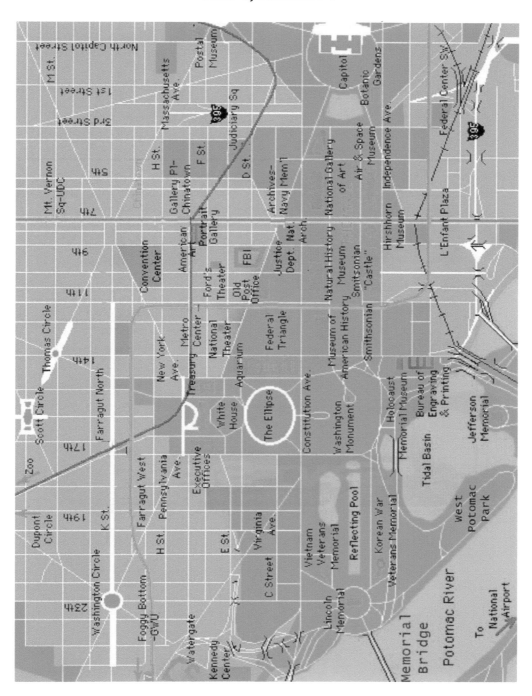

Topographic

Railroad—single track: multiple track:		Primary highway: Dual highway:	
Bridge:		Secondary highway:	
Buildings (dwelling, place of employment, etc.):		Light-duty road:	
		Unimproved road:	
Buildings (barn, warehouse, etc.):		Trail:	
School:		Topographic symbols— Index contour:	
Church:		Intermediate contour:	
Cemetery:		Supplementary contour:	
Tanks—oil, water, etc. (labeled only if water):	Water Tank	Depression contour:	
		Checked spot elevation:	x 5970

Purple indicates revisions from aerial photographs and other sources; not field checked.

SPACE STARS™
Laboratory for GIS/Remote Sensing Education

Source: U.S. Geological Survey

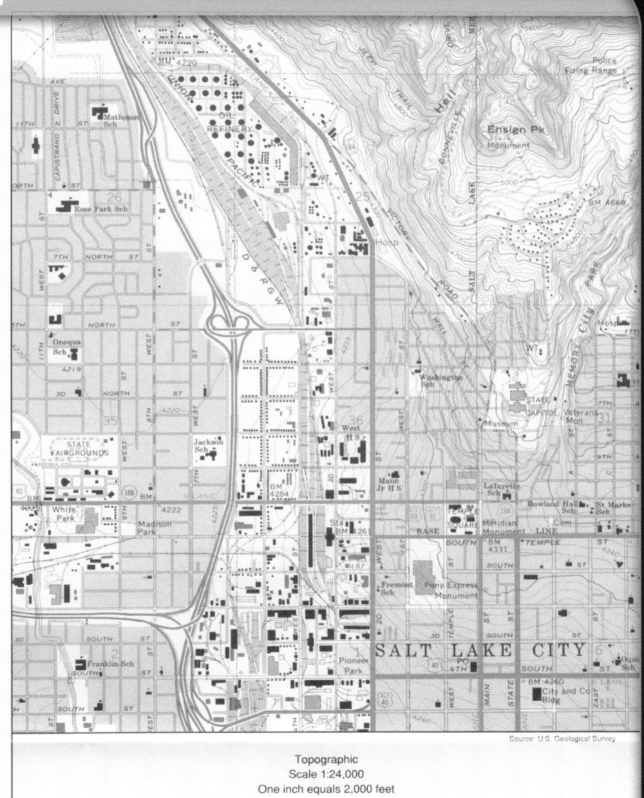

Source: U.S. Geological Survey

Topographic
Scale 1:24,000
One inch equals 2,000 feet

Contour interval 20 feet

State capitol:		Airport:	✈
School:		Route markers—	
Hospital:		interstate; U.S.; State:	

Population key

● SALT LAKE CITY	100,000 to 500,000	● Draper	5,000 to 25,000
◉ Ogden (county seat)	50,000 to 100,000	◉ Heber City (county seat)	1,000 to 5,000
● Murray	25,000 to 50,000	● Henefer	500 to 1,000
		● Daniels	up to 500

SPACESTARS™
Laboratory for GIS/Remote Sensing Education

Source: U.S. Geological Survey

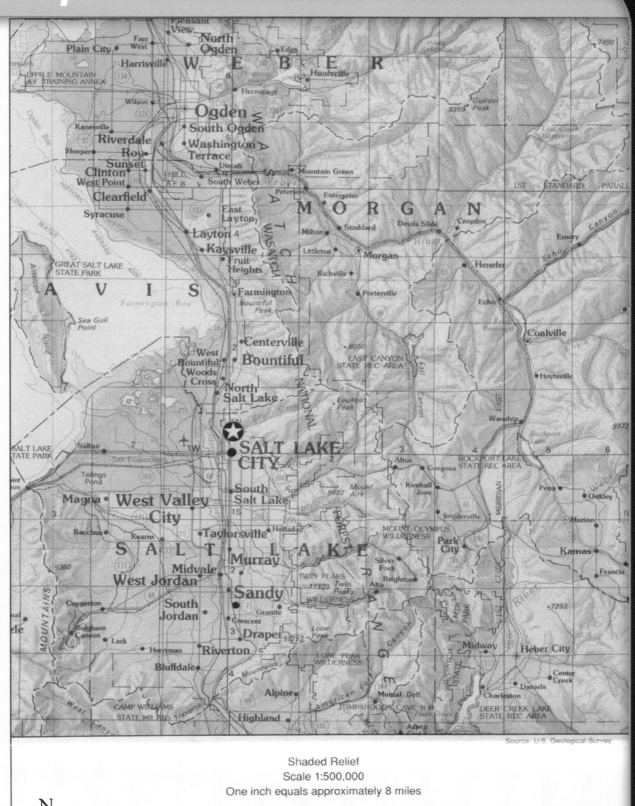

Source: U.S. Geological Survey

Shaded Relief

Scale 1:500,000

One inch equals approximately 8 miles

N

Contour interval 500 Feet

State capitol:	County boundary:	
Interstate highway:	Route markers—	
Multilane divided highway:	interstate; U.S.; State:	
Principal through highway:	Built up area; locality; elevation:	Elev. 5970
Other highway:	Airport:	
Local road:	Campsite:	
	Rest area:	

Paved Improved Unimproved

Paved Improved Unimproved

Population key

SALT LAKE CITY	100,000 to 500,000
OGDEN	50,000 to 100,000
Milton	500 to 1,000

Source: Utah Department of Transportation

SPACESTARS
Laboratory for GIS/Remote Sensing Education

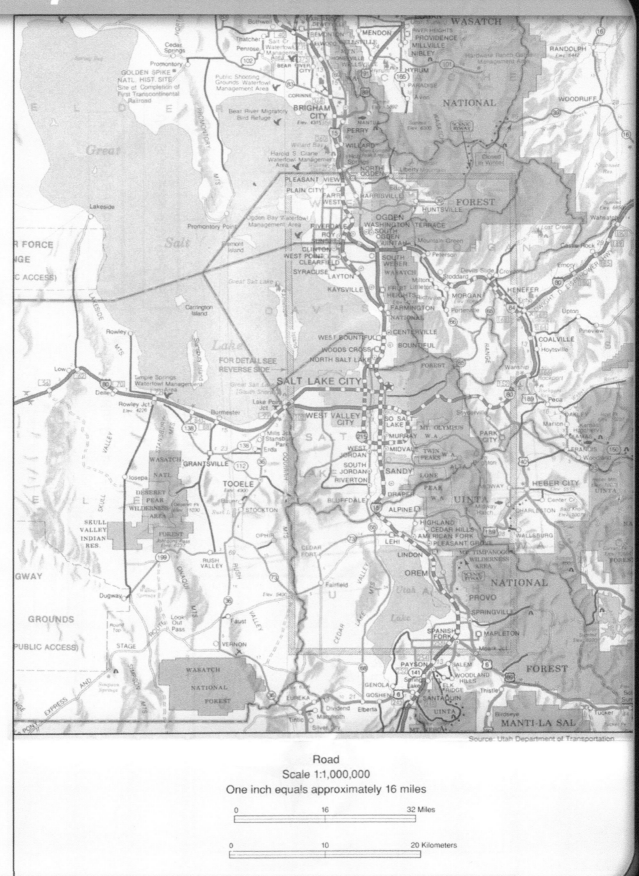

Source: Utah Department of Transportation

Road

Scale 1:1,000,000

One inch equals approximately 16 miles

0   16   32 Miles

0   10   20 Kilometers

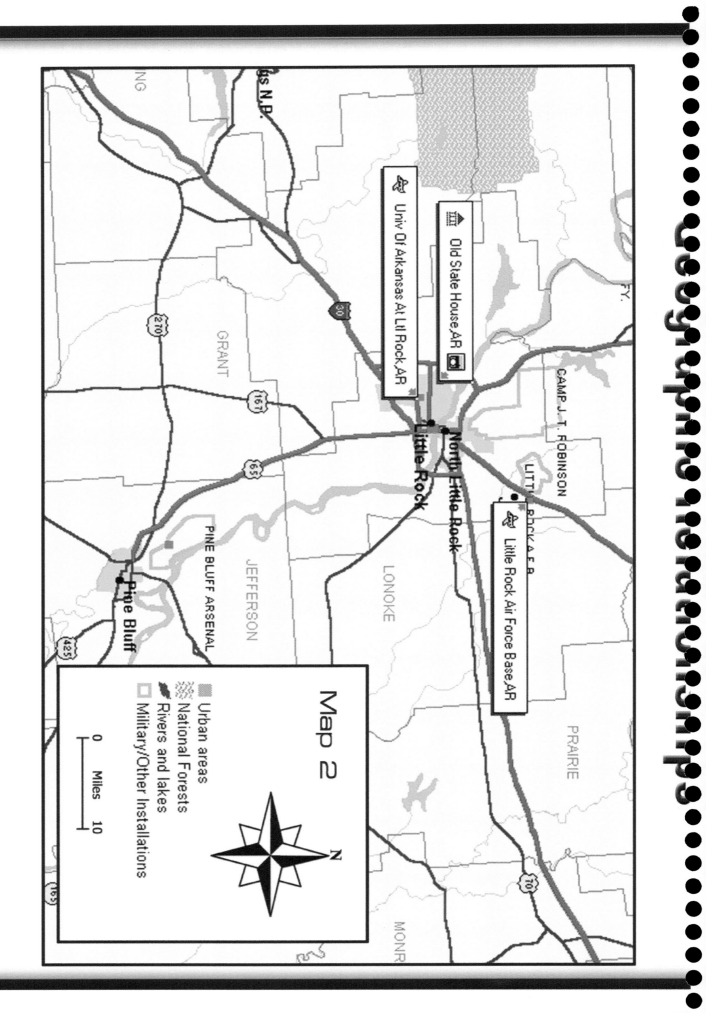

Map 2

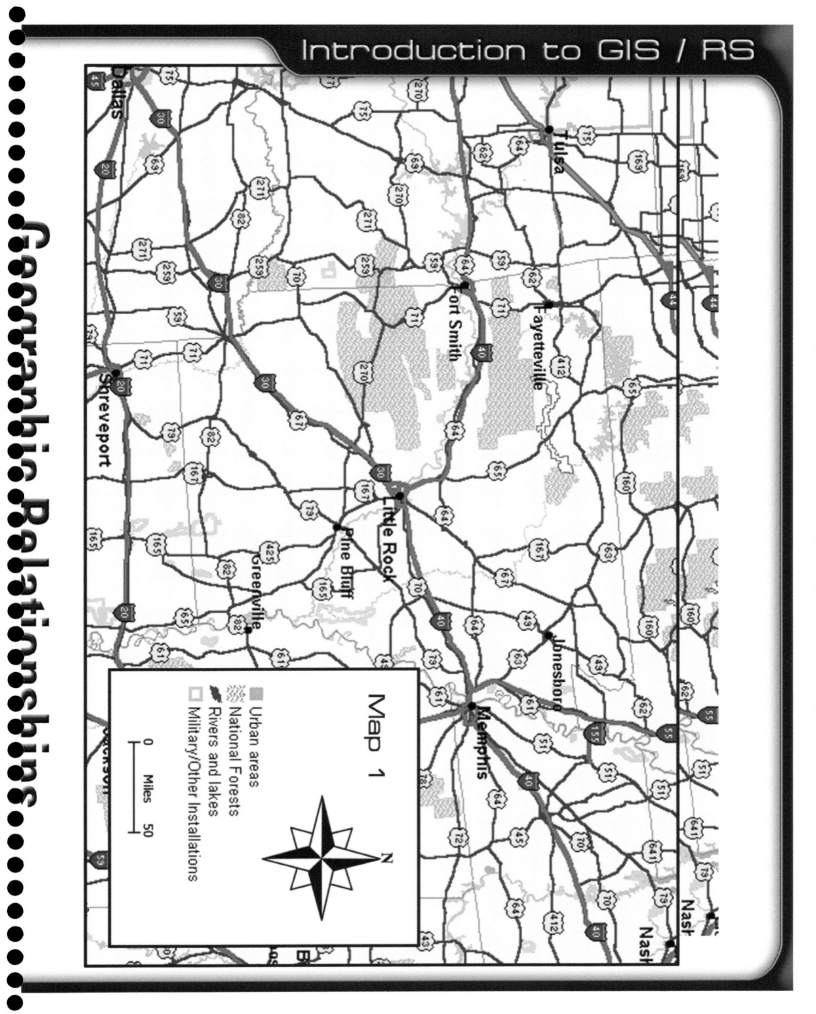

Geographic Relationships

Glossary

Absolute Location: can be used on 2D or 3D objects, acquired by using a coordinate system to determine a referenced, fixed point in space

Active Remote Sensing: uses electronically generated signals or light waves that bounce off targets

Aerial Images: images taken from aircraft that provide a detailed view of the landscape

AOI – Area of Interest : defines the area that in which project research is conducted

Attributes: descriptive information included in tables that are included with every point, line, and polygon layer in a GIS

Azimuthal Projection: a projection which is planar and is created as if a flat piece of paper (a plane) is placed touching the globe and the geographic features on the globe are projected onto the paper

Bar Scale: is a line or bar that has tick marks for units of distance

Cartographers: the people who make those maps

Cartographic Process: the process of making a map includes four steps; data collection, data compilation, map production, and map reproduction

Cartography: the science or art of making maps

Conic Projection: computed as if a piece of paper is shaped as a cone and placed around the globe

Continuous Data: data that does not have distinct boundaries – oftentimes used to describe raster data.

Coordinate Reference Systems: another term for a way to communicate where something is located on the surface of the Earth using an x,y coordinate

Cultural Maps: shows patterns of demographic data

Glossary

Cylindrical Projection: projection that simulates a piece of paper wrapped around the globe

Data Acquisition: getting information such as shapefiles or image files from credible sources whether from the Internet or from a private company

Data Collection: the act of gathering data and putting it in a usable format, may also involve hiring a specialist to collect data for a study

Data Layer: digital layer of information about one entity in a map

Data Processing & Analysis: taking the data collected and plugging it into a geospatial software program where it can be analyzed

Discrete Data: data that represents phenomena with distinct boundaries – used to describe vector data

Eastern Hemisphere: the area located east of the Prime Meridian to 180° longitude

Easting: used in both the UTM and State Plane systems, this represents the "x" value providing the measurement east of the point of origin.

Electromagnetic Spectrum: is a scale that displays the arrangement of electromagnetic radiation in terms of energy, wavelength, or frequency.

Ellipsoid: shape used to describe the shape of the earth – a spheroid that is slightly flattened, making it appear wider in the center portion

Equator: starting point for the lines of latitude is at the horizontal center of the earth, dividing it into the northern and southern hemispheres

False Color Image: an imaging technique that does not use traditional true color – used in remote sensing for image enhancement purposes.

Functional Requirements: predicting what you need to conduct your study like data, equipment, computer hardware and software, personnel, and supplies

Geodetic Datum: mathematical computations that describe the irregular shape of the Earth

Geographic Coordinate System: establishes two sets of imaginary lines around the earth: parallels, representing latitude, and meridians, representing longitude, forming a "grid" over the earth

Geographic Extent: the area represented by the map/data frame; defined by a longitude along the sides and latitude on top and bottom

Geographic Information Systems - (GIS): is a combination of data, computer software and hardware system for storing, sorting, processing, and displaying geographic information

Heads-up Digitizing: a method of collecting vector (point, line, and polygon) data by tracing or marking elements in a GIS and storing them digitally

High Resolution Imagery: imagery with cell sizes that represent a smaler area; used in large scale maps

Horizontal Geodetic Datum: datums that are used calculate positions on the Earth's surface

International Date Line: imaginary line that separates two calendar days located approximately 180° from the Prime Meridian

Large Scale Map: a map that displays a smaller area with features shown in greater detail

Latitude: lines that run east and west along the earth's surface, because they do not touch are also called parallels – measured from 0° at the equator to 90° at the poles

Longitude: lines that run north and south along the earth's surface, also called meridians, these lines do intersect at the poles – measured from 0° at the Prime Meridian to 180°

Low Resolution Imagery: imagery with cell sizes that represent a larger area; used in small scale maps

Map: visual representation of an area shown on a two dimensional surface

Map Projection: a two dimensional representation of the Earth's surface – it will include some distortions in area, shape, direction, or distance

Map Scale: the relationship between distance on the map and the portion of the earth that it represents

---

Glossary

Meridians: term used to describe lines of longitude

Metadata: information about data. In ArcMap, it is a file that is attached to a shapefile or image to show its source, the content, geographic extent, and other details about the data file.

Multispectral Imagery: imagery that is composed of a combination of several bands within the electromagnetic spectrum; can be used to enhance the ability to see features that may not be detectable

Northing: used in both the UTM and State Plane systems, this represents the "y" value providing the measurement north of the point of origin.

Oral Presentation: delivering communication for a study directly to stakeholders relating information such as project findings, project summary, information on project implementation, any deviations from original plan, and conclusions or recommendations for further study.

Orthogonal Viewpoint: viewing something from above

Parallels: another term used to describe lines of latitude

Passive Remote Sensing: uses sensors or cameras that produce imagery gathered via energy or light that is naturally reflected from objects on the ground

Physical Maps: shows land and water formations on the Earth's surface

Political Maps: shows how humans have divided the Earth

Prime Meridian: measured at 0° longitude, is the starting point of the meridians

Problem Identification: the act of defining the question to be considered, solved, or answered

Project Design: involves creating an outline of the entire project management process

Project Feasibility: determined in terms of not only monetary costs, but also in terms of time, personnel, and material costs involved versus the potential benefits that will be received from conducting the study

Project Implementation: the second phase of the PMM, involves acquiring the resources and data, processing the data, and creating maps in order to answer geospatial related project questions

Project Objective: establishes the project's goals so that all personnel know the focus and scope of the project

Project Planning: the first phase in the Project Management Model, a list of tasks that will provide the foundation for the rest of the project

Project Presentation: communicating the findings of a study to the stakeholders.

Project Stakeholders: individuals, families, agencies, or organizations that have a keen interest in the project and/or the project findings or outcomes

Project Title: a single, concise idea for all of the coordinated activities involved in the project

Raster Data: continuous data that is made of equally sized cells that have unique numeric values. Each cell represents a specified area on the ground. Raster data is often created from sample data or remotely sensed data (such as imagery).

Relative Location: a way of describing where an object is on the earth's surface without giving absolute coordinates

Remote Sensing: refers to gathering information about our world from a distance

Representative Fraction: a type of scale that is shown in ratio format

Resource Acquisition: collecting and acquiring the functional requirements that were identified during the Project Planning phase of the PMM

Satellite Imagery: an image taken from a satellite; able to cover a large geographic area

Small Scale Map: a map that shows a larger area with less detail

Spatial Reference Systems: term for a way to communicate where something is located on the surface of the Earth using an x,y coordinate

Glossary

State Plane: a coordinate system that is used exclusively in the United States. This coordinate system focuses in on a zone or zones within a state. Just as with UTM, measurements are made in feet using Easting for "X" coordinates and Northing for "Y" coordinates.

Tasked: ordering an image that has not yet been acquired by any organization

Thematic Maps: maps created that show a specific theme or use

Tropic of Cancer: located at 23°30' N latitude divides the tropics from the northern temperature zone

Tropic of Capricorn: located at 23°30' S latitude divides the tropics from the southern temperature zone

Tropics: the area between the Tropic of Cancer and Tropic of Capricorn, experiences no dramatic changes in seasons within this zone because the sun is always directly overhead

True Color Image: An image that appears with colors that the human eye can detect.

UTM: Universal Transverse Mercator - grid system originally set up for military purposes splits up the world into 60 zones, 6° wide. The units of measurement for UTM are meters.

Vector data: data displayed as points, lines, or polygons. Each has a specific longitude and latitude coordinate or group of coordinates to describe their location on the Earth and each has a specific value applied to that shape.

Verbal Scale: explains the scale in words

Vertical Geodetic Datum: datums used to calculate heights from the Earth's surface

Western Hemisphere: the area located west of the Prime Meridian to 180° longitude

Written Report: a written document listing all details of a study including project findings and analysis, any project deviations, and recommendations for future study